Principles of Soil Chemistry

BOOKS IN SOILS, PLANTS, AND THE ENVIRONMENT

Additional Volumes in Preparation

Principles of Soil Chemistry

Second Edition

Kim H. Tan

Department of Agronomy
The University of Georgia
Athens, Georgia

Marcel Dekker, Inc.　　　**New York • Basel • Hong Kong**

Library of Congress Cataloging-in-Publication Data

Tan, Kim H. (Kim Howard)
  Principles of soil chemistry / Kim H. Tan -- 2nd ed.
    p.   cm. -- (Books in soils, plants, and the environment)
  Includes bibliographical references (p.   ) and index.
  ISBN 0-8247-8989-X
  1. Soil chemistry.   I. Title.   II. Series.
  S592.5.T36   1992
  631.4'1--dc20   92-35625
                                                          CIP

This book is printed on acid-free paper.

MARCEL DEKKER, INC.
270 Madison Avenue, New York, New York 10016

Current printing (last digit):
10 9 8 7 6 5 4 3 2 1

PRINTED IN THE UNITED STATES OF AMERICA

*Dedicated to the memory of
Dr. Ir. J. Van Schuylenborgh,
the author's mentor and
deacon of the subfaculty of the
University of Amsterdam, Amsterdam,
The Netherlands*

Preface to the Second Edition

The first edition of *Principles of Soil Chemistry* was tested as a textbook for more than ten years at the University of Georgia and it received favorable comments from a great number of undergraduate and graduate students in my soil chemistry classes. It has been used by other universities in the United States, and it has been adopted as a textbook by universities in other countries (for example, Morocco, Egypt, Jordan, Hungary, Venezuela, Argentina, and Indonesia). The book has been translated into French, Spanish, Hungarian, and Indonesian. Many scientists and professionals who work in consulting, in private analytical laboratories, and in the fertilizer and humic acid industries have indicated how useful this book is.

A second edition was thus warranted, with an updated text that reflects the most recent advances in soil chemistry and in environmental quality issues related to soils. The organization of the text in the new edition has changed somewhat, but the principle remains the same: organizing the book in the context of a soil system composed of gas, liquid, and solid phases. The first chapter, on the basic chemical and thermodynamic principles, has been enlarged. A section is included on the modern concept of elementary particles—the basic components of matter in the universe—advanced today by nuclear physicists. The problem of electron activity has been added in Chapter 2, and topics on soil air and water, the soil gas and liquid phase, are discussed in Chapter 3. New sections are included in Chapter 3, one of which is related to environmental quality and pollution of soil water.

Chapter 4 of the first edition has been divided into Chapters 4 and 5 to incorporate many changes in organic chemistry and soil mineralogy. All organics are now treated in Chapter 4, and among the new materials discussed is the importance of complex carbohydrates in the environment. The discussion of lignin and lignification has been enlarged considerably. The section on humic acids has been revised to incorporate the many new advances in humic acid science. The agricultural, industrial, and ecological importance of humic matter are brought to the foreground. Chapter 5 deals only with the inorganic soil constituents, and it contains new information on soil minerals and soil clays. New concepts about noncrystalline clays are included, and I have added a description of the electric triple layer as the newest model in surface chemistry of clays.

Basic chemical reactions of the soil components are discussed in the Chapters 6 through 11. The adsorption theories have been enlarged in Chapter 6, and a new concept, osmotic adjustment, has been added to explain adsorption of water by plants under duress. Cation exchange theories have been redefined in Chapter 7 to increase comprehension. Anion exchange and phosphate fixation are now discussed in Chapter 8, and the phosphate potential has been redefined to include the chemical and electrochemical potential of phosphate. Chapter 9 now covers soil reactions, and the role of aluminum chemistry contributing to soil-acidity is discussed in detail. The role of fertilizers, pyrite, and mine-spoil in soil acidity are included, and the issue of acid rain, contributing to soil acidity and declining environmental quality, is assessed. Chapter 10 gives an enlarged presentation of the role of soil chemistry in soil formation, the importance of the redox potential in soil formation, and fertility. The application of soil chemistry in soil-organic matter interaction is now discussed in Chapter 11, the final chapter. The importance of soil organic acids and their chelation reactions are discussed more explicitly.

I want to thank Dr. J. B. Jones, Jr., Director of M.M.I., Inc., Athens, Georgia, and Dr. Harry A. Mills, Professor of Horticulture, University of Georgia, for their encouragement and constructive criticism. My appreciation is extended to Dr. D. L. Sparks, Professor and Chairman, Department of Plant and Soil Sciences, University of Delaware, for his critical but objective review of the first edition, which made this revision possible; and to Dr. M. G. M. Bruggenwert, co-editor of *Soil Chemistry: A. Basic Elements* (Elsevier, The Netherlands), and Dr. Patrick MacCarthy, Department of Chemistry and Geochemistry, Colorado School of Mines, for being so generous with criticisms. Thanks are extended to the anonymous reviewers and many other unnamed people, who have assisted in the development of the second edition. Thanks are also due to the various publishers, societies, journals, and fellow scientists who gave permission to reproduce figures and diagrams; and to Mr. John A. Rema and Mr. Budi Irianto for their valuable assistance with the laser printer in the development of the draft of the second edition. Finally, I want to acknowledge the encouragement and understanding of my wife, Yelli, and my son, Budi, during the progress of this revision.

Kim H. Tan

Preface to the First Edition

This book presents the adaptation of pure chemical science to the scientific study of soils and plants. The text provides comprehensive coverage of the fundamental topics in soil chemistry. Its unique approach, including a definite soils' flavor by the integration of organic and inorganic components in the dynamic processes in soils, important for continuation of life, is not currently available. In plain language, easy to comprehend by a wide range of scholars and students, but without sacrificing the scientific value, the book starts with a review of basic chemical principles and thermodynamics pertinent to the following topics on concepts and processes in and related to the soil solution: colloidal organic and inorganic components, their modern classification, and reactions and interactions affecting changes in the behavior of soils and plant growth. The book tells you every thing that you want to know about humic acids. Separate chapters are included on the use of x-ray diffraction, infrared analysis, differential thermal analysis (DTA), and other methods for the identification of organic and inorganic soil constituents. Examples are given in interpretation of results using tables provided on diagnostic d spacings in x-ray analysis, wave numbers in infrared spectroscopy, and peak temperatures in DTA. Crystal chemistry of inorganic compounds and surface chemistry of inorganic and organic colloids are explained in simple terms, showing their significance in the control of the many complex reactions in nature. The traditional adsorption, cation and anion exchange, solid acidity, and salinity theories are presented together with the current concepts. In the final two chapters, the principles of soil chemistry are applied in soil formation and in soil–organic matter interaction. Although the purpose of the book is to fill a need in soil science, that is, by bridging pure chemistry and soil science, the volume is equally useful in explaining the soil as a basic entity for related disciplines in agriculture and other sciences, for example, crop and plant science, irrigation, forestry, conservation, plant physiology, ecology, microbiology, geology, geochemistry, physics, chemistry, and botany.

Special recognition goes to Dr. Elvis R. Beaty, Professor of Agronomy, University of Georgia, for editing the book. My thanks are also due to Dr. Ralph A. Leonard, Research Leader, Southeast Watershed Research Program, USDA, -SEA, -AR, and to Dr. Robert A. McCreery, Associate Professor of Agronomy, University of Georgia, for reading the manuscript for correct English usage and scientific value. Appreciation is extended to Dr. J. B. Jones, Jr. Former Director, Soil Testing Laboratory, Cooperative Extension Service, and Former Division Chairman, Department of Horticulture, University of Georgia, for his valuable criticism, and to the unnamed people who have assisted in the development of the book. Thanks are also extended to the various publishers, scientific societies, and fellow scientists, who gave permission to reproduce figures, photographs, and diagrams. Finally, the author wants to thank his wife, Yelli, and his son, Budi, who always stood by with great enthusiasm and a lot of encouragement.

Kim H. Tan

Contents

Principles
of
Soil
Chemistry

1

Review of Basic Chemical Principles

1.1 ATOM AND ATOMIC STRUCTURE

An *atom* is the smallest particle of an element that can enter into a chemical combination. Atoms of the same elements are similar in composition, but one element differs from the other in size, position, and movement of its atoms. An *element* is a substance composed of atoms with the same atomic number, or nuclear charge. In solid matter, the atoms vibrate within the confines of very small spaces, whereas in gas the atoms exhibit a considerable range of movement.

The concept of atoms being the smallest particles of matter was first postulated by Democritus or Leucippus, Greek philosophers, in approximately 425 B.C. The term atom comes from the Greek *atomos*, meaning indivisible. However, it was not until Dalton's atomic theory was formulated in the first decade of the 19th century, that this idea became scientifically established. Although John Dalton (1808) is generally credited as the founder of the atomic theory, he was only reviving the old Greek atomic hypothesis and gave it scientific credibility. Since then, Crookes, Thomson, and others, working on the conduction of electricity in rarified gases, made revisions in this theory and concluded that the atom was composed of still smaller particles. The structure of the atom became a subject of research interest, and by the end of the 19th century it became known that the atom had the following components:

1. *Electrons*, small negatively charged components of atoms of all substances.
2. *Protons*, positively charged particles of much greater mass than electrons

With the advancement of science in the 20th century, it became clear that atoms also contain neutrons. The *neutrons* have a mass number of 1, but have zero (0) charge. Less fundamental particles were also detected, the positrons. *Positrons* are particles with the mass of an electron and the charge of a proton. Currently, several other subatomic particles have been detected or are produced in atomic "smashers" or "accelerators" (Applequist, 1986; Glashow, 1986; Orr, 1986; Gaillard, 1983; McCarthy et al., 1983). By pounding or pummeling the atom with fast-moving subatomic particles in an atomic accelerator, the atom's central part, the *nucleus*, is torn apart into smaller particles. The smaller the particles to be obtained, however, the more powerful and costlier the machines should be. Depending on the energy created and the production of an ever increasingly smaller division of particles, the atom accelerators can be distinguished into:

1. Megatrons, capable of accelerating protons to 1 mega (million) electron volts (eV; 1 MeV $= 1 \times 10^6$ eV).
2. Gigatrons, capable of accelerating protons to 1 giga (billion) electron volts (eV; 1 GeV $= 1 \times 10^9$ eV).
3. Tevatrons or teratrons, the largest and highest-energy proton synchrotron in the world. It is capable of accelerating protons to 1 tera (trillion) electron volts (eV; 1 TeV $= 1 \times 10^{12}$ eV).

At the Fermi National Accelerator Laboratory in Chicago, Illinois, nuclear physicists can literally tear apart the fabric of matter with its Tevatron. Fermilab's European rival, Centre of European Nuclear Research (CERN), located in Geneva, Switzerland, has discovered the building blocks of neutrons and protons. These smallest particles on earth are now called *elementary particles*, which, in fact, include both atomic and nonatomic components. Four major groups

of elementary particles can be recognized (e.g., quarks, leptons, neutrinos and force carriers).

Quarks

Protons and neutrons are traditionally considered the elementary
constituents of the atomic nucleus. They are collectively called *baryons*. However, it is now established beyond doubt that all baryons
can be broken down into more elementary constituents, the quarks.
Protons, as now understood, are not elementary particles, but rather
bags, containing quarks and gluons. They are the heavy components
in atoms. Many types of quarks have been identified (e.g., b, c, d,
and t quarks).

Leptons

Electrons are the best known *leptons*. They are the light components
of atoms, and are less numerous in type than quarks. Both quarks
and leptons are pointlike particles of spin $0.5h/2\pi$ (where h = Max
Planck's constant) and are considered the constituents of all matter.
Recently, the Stanford Linear Accelerator Center (SLAC) in California has detected another kind of lepton, the "Tau lepton."

Neutrinos

In contrast with electrons, which are charged leptons, *neutrinos* are
neutral leptons. Three types of neutral leptons, collectively called
neutrinos, have been discovered in cosmic rays, in the form of waste
energy from nuclear reactors, in the form of decay products of radioactive material, and as intense beams produced by high-energy
particle acceleration. Physically, neutrinos are not components of
atoms or molecules, but are force carriers like photons. Neutrinos
play an important role in the universe and are essential to the nuclear fusion that powers the sun. They are important ingredients in
the "beta-process," by which hydrogen can be transmuted into various elements found on earth. In the absence of neutrinos, the earth
would exist as a solid ball of frozen hydrogen.

Force Carriers

Photons are field quanta or force carriers. They are massless elementary excitation particles of the electromagnetic field, and physically are not components of the atoms. Many types of photons can be distinguished (e.g., photons from microwaves, photons from infrared radiation, photons from ultraviolet waves, and others).

Although much more is known currently on subatomic components, the atomic model, as proposed by Rutherford and Bohr (Cragg, 1971; Garrett, 1962; Hall, 1966), is still prevalent today. It represents the modern idea of the atomic structure, which provides a reasonable explanation for many phenomena in atomic physics, chemical reactions, and valencies. According to this model, the fundamental particles of the atom are (1) electrons, (2) protons, and (3) neutrons. Protons and neutrons are located in a small central portion of the atom called the *nucleus*. The nucleus is of high specific weight and contains most of the mass of the entire atom. The mass of electrons is very small and can be neglected. The various groups of electrons are placed in concentric shells around the nucleus. The shells may be composed again of subshells or cells. Neglecting the presence of subshells, the shells may contain one electron, as in the hydrogen atom, or two electrons as in the helium atom, or more (Figure 1.1).

The first shell adjacent to the nucleus is called the *K shell*, whereas the shell next to it is designated as the *L shell*, and so on. The most dense atom is the uranium atom (^{238}U), with 92 electrons distributed around the nucleus in K, L, M, N, O, P, and Q shells. The diameter of the nucleus is between 1×10^{-13} and 1×10^{-12} cm. Whereas the nucleus carries an integral number of positive charges, or integral number of protons, each of 1.6×10^{-19} C, each electron carries one negative charge of 1.6×10^{-19} C. This value is generally known as the *electron charge* (e).

The electrons in the inner shells are tightly bonded to the nucleus. This inner structure can be altered by bombardment with high-energy particles, and the resulting excitation of electrons in the K shell may result in emission of Kα, and Kβ radiation, called *x-rays*, or *roentgen radiation*. With most atoms, it is the arrangement of energy in the outer shells that undergoes changes during chemical reactions. These outer-shell electrons, called *valence electrons*, are responsible for the chemical properties of the atom. During these

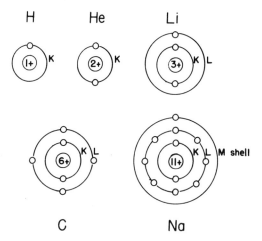

Figure 1.1 Atomic structure showing the K, L, and M electron shells.

changes, the role of the nucleus is usually passive. The hydrogen atom is perhaps an exception, since it has only one shell and one bare proton.

Since an atom is electrically neutral, the total positive charge in the nucleus must equal the total negative charge of the electrons. An atom that loses one or more electrons from the outer shell assumes then a net positive charge, and is called a *cation* (Faraday). When an atom has excess electrons, not balanced by the positive charges of the nucleus, it assumes a net negative charge and is called an *anion*.

1.2 ATOMIC MASSES AND WEIGHTS

Mass

The term *mass* is defined as the quantity of matter contained in a body. The *force* that gives the body a given acceleration is a measure of mass.

Weight

Weight, on the other hand, pertains to the force of attraction of the earth for the body. This attraction is dependent on the distance the body is from the center of the earth. On earth, mass = weight, since the body is close to the center of gravity.

In soil chemistry, the terms *atomic mass, atomic mass number, atomic number*, and *atomic weight* are frequently used.

Atomic Mass and Atomic Mass Number

The *atomic mass* is the sum of the masses of protons, neutrons, and electrons and, therefore, is the exact mass of an atom. However, as indicated before, the masses of electrons are extremely small and can be neglected. Therefore, the *atomic mass number* (symbol A) has been proposed to represent the sum of the masses of protons and neutrons only. It is the nearest integer of the atomic mass or exact mass. In other words, the exact value is called the atomic mass, whereas the nearest whole number of the atomic mass is referred to as the atomic mass number.

Atomic Mass Unit

Atomic masses can be expressed on a chemical or a physical scale. In general, masses on the chemical scale are smaller than a weighted average physical mass by a factor 0.999973. On a chemical scale the element oxygen is arbitrarily assigned the mass 16, nitrogen the mass 14, carbon the mass 12, hydrogen the mass 1, and so on. Weight units can be assigned to the atomic masses as follows:

gram atom = mass of an element in grams

According to this definition 1 gram atom of oxygen is then 16 g, 1 gram atom of nitrogen is 14 g, and 1 gram atom of carbon is 12 g.

However, the unit mass of single atoms is usually expressed in terms of atomic mass units (amu). *One atomic mass unit* (1 amu) is, by definition, $\frac{1}{12}$ of the mass of the most abundant isotope of carbon, (^{12}C);

1 amu = $\frac{1}{12}$ of ^{12}C = 1.66 × 10^{-24} g

In some books the unit amu is also called *unified atomic mass constant*.

Mass is related to energy by the famous equation as formulated by Einstein:

$$E = mc^2$$

in which E = energy in ergs, m = mass in grams amu, and c = velocity of light in vacuum (3×10^{10}cm/sec).

Atomic Weights

Strictly speaking, *atomic weights* are not weights at all. They are simply numbers that indicate the relative weights of the different kinds of atoms, and no reference is made to absolute weights. Hydrogen was originally assigned a relative weight of 1, since it was considered the fundamental particle and the lightest of all atoms. The heaviest of any of the naturally occurring atoms is uranium with an atomic weight of 238. When we say that the element oxygen has an atomic weight of 16, we simply indicate that the oxygen atom is 16 times heavier than the hydrogen atom. Therefore, no weight units have been assigned to the numbers. As with atomic masses, however, weight units can be assigned as follows:

gram atomic weight = atomic weight in grams

The mass of an element in grams is numerically equal to the atomic weight in grams. Therefore, 1 gram atom of oxygen = 1 gram atomic weight of oxygen = 16 g, and 1 gram atom of carbon = 1 gram atomic weight of carbon = 12 g.

Atomic weights (AW) apply only to elements. Compounds have *molecular weights* (MW), which equal the sum of the atomic weights of the elements making up the compounds.

The selection of hydrogen as a standard base for comparative assessment of other atomic weights was later changed in favor of oxygen, which is assigned an arbitrary mass of 16.0000. The atomic weight of hydrogen changes consequently to 1.0080. To comply with the suggestion from atomic physicists, this standard base was revised again in 1961, and atomic weights are currently based on the assigned relative mass of ^{12}C (AW = 12.0000). However, the consequent changes in atomic weights of the other elements are very small (Int. Union of Pure and Applied Chemistry, 1974).

Table 1.1 Absolute Weights of One Atom (or One Molecule) of Selected Elements

Substance	Mole	AW or MW	Grams	Number of particles or atoms	Weight of one atom
H^+ ion	1	1	1	6×10^{23}	1.67×10^{-24}
Carbon	1	12	12	6×10^{23}	2.0×10^{-23}
Na	1	23	23	6×10^{23}	3.83×10^{-23}
K	1	39	39	6×10^{23}	6.50×10^{-23}
Ca	1	40	40	6×10^{23}	6.67×10^{-23}
NaCl	1	58	58	6×10^{23}	9.67×10^{-23}
KCl	1	74	74	6×10^{23}	1.23×10^{-22}
$CaCO_3$	1	100	100	6×10^{23}	1.67×10^{-22}
$C_6H_{12}O_6$(glucose)	1	180	180	6×10^{23}	3.00×10^{-22}

1.3 AVOGADRO'S NUMBER

The number of atoms in 1 gram atomic weight of any element is

$$6 \times 10^{23}$$

This number is known as the *Avogadro's number* (Table 1.1).

How much does one atom of hydrogen weigh? Since 1 gram atom of hydrogen weighs 1 g, and since 1 gram atom of hydrogen contains 6×10^{23} atoms, then one atom of hydrogen will weigh $\frac{1}{6} \times 10^{23} = 1.67 \times 10^{-24}$ g, an extremely small number. Even the actual weight of an uranium atom is inconceivably small. Yet, it is 238 times heavier than is a hydrogen atom.

1.4 ATOMIC NUMBER

The *atomic number* is the number of protons and, hence, the nuclear charge (symbol Z), in the atomic nucleus. Then the number of neutrons equals the difference between the atomic weight and the atomic number Z.

1.5 ATOMIC ORBITALS

Atomic orbitals are functions that define the spatial behavior of electrons of given energy levels in a particular atom. The number of orbital electrons is equal to the number of protons in the nucleus, so that the atom as a whole has a net charge of zero.

1.6 ATOMIC RADIUS

The electron density around an isolated atom extends to infinity. However, if one is referring to the size of an atom in a molecule or a crystal, then one may define the *atomic radius* as the closest distance of approach of a probe to the nucleus (Table 1.2).

As we will see later on, the size of elements will play an important role in many soil chemical reactions. Cation exchange reactions, hydration, double layers, and problems with potentials, dispersion, and flocculation of colloids, and so on, all are affected by the size of ions in the reactions.

1.7 VALENCE

The *valence* of an atom or element is that property which is measured by the number of atoms of hydrogen (or its equivalent) that can be

Table 1.2 Ionic Radii, Atomic Numbers, and Atomic Weights of Selected Elements

Ion	Radius (Å)[a] Crystalline	Hydrated	Atomic number	Atomic weight
Si^{4+}	0.42		14	28.09
Al^{3+}	0.51	9	13	26.98
Fe^{2+}	0.74		26	55.84
Fe^{3+}	0.64	9	26	55.84
Ca^{2+}	0.99	4.3	20	40.08
Mg^{2+}	0.66	4.2	12	24.32
Ba^{2+}	1.35	4	56	137.36
Li^{+}	0.60	3.8	3	6.94
Na^{+}	0.98	3.6	11	22.99
K^{+}	1.33	3.3	19	39.10
Rb^{+}	1.48	3.3	37	85.48

[a] 1 Å (angstrom) = 1×10^{-8} cm or 1 Å = 0.1 nm.
Source: Weast (1972) and Gast (1977).

held in combination by one atom of that element if negative, or that can be displaced by one atom of the element if it is positive. In simple terms, the valence is a measure of the combining capacity of atoms. Atoms with the smallest combining capacity are considered to have a valence of 1. Valences are whole (integral) numbers and correspond to the number of valence electrons the atom carries. Valence electrons are electrons which are gained, lost, or shared in chemical reactions.

1.8 EQUIVALENT WEIGHT

The *equivalent weight* (Eqw) of an atom or ion is defined as

$$\frac{\text{Atomic (Formula) Weight}}{\text{Valence}}$$

If the atomic weight is assigned as gram atomic weight, the equivalent weight is then in grams. If the atomic weight is in milligrams (mg), then the equivalent weight is in milligrams.

Elements entering into a reaction always do so in amounts proportional to their equivalent weights. The value of the atomic weight does not change with the reaction, but the valence of the atom will change depending on the (1) type of analysis, and (2) the reaction process. The following examples for the determination of equivalent weights serve as illustrations:

Acid–Base Titrations

1. In this type of analysis involving reactions with monovalent ions, 1 equivalent (Eq) equals 1 mole (mol).
2. With polyvalent ions, the equivalent weight is variable and depends on the reaction process:

$$H_3PO_4 \rightarrow H^+ + H_2PO_4^-\qquad 1\ Eq = 1\ mol$$
$$H_3PO_4 \rightarrow 2H^+ + HPO^{2-}\qquad 1\ Eq = \tfrac{1}{2}\ mol$$
$$H_3PO_4 \rightarrow 3H^+ + PO^{3-}\qquad 1\ Eq = \tfrac{1}{3}\ mol$$

Precipitation and Complex Reactions

In this type of analysis, the relationship between the equivalent weight and numbers of moles can be read directly from the reaction.

In volumetric analysis by which cyanide is titrated with silver according to the Mohr method, the reaction occurs as follows:

$$Ag^+ + CN^- \rightarrow AgCN$$

Here the equivalent equals 1 mol. If the Liebig method is used in the titration of cyanide with silver, the end point of titration is reached when the following reaction has occurred:

$$Ag^+ + 2CN^- \rightarrow Ag(CN)_2^-$$

Consequently, the equivalent of cyanide equals 2 mol.

Oxidation–Reduction Reactions

In these reactions, the *equivalence* of a substance is, by definition, that part of a mole that in its reaction corresponds to the removal of $\frac{1}{2}$ gram atom of oxygen, or the combination with 1 gram atom of hydrogen or any other univalent element. One way to find the equivalent is to determine in the reaction the change in oxidation state of the element, as shown in the following examples:

1. In the titration of ferrous into ferric iron using an oxidation agent, the state of oxidation of iron changes from 2 to 3:

   $$Fe^{2+} \rightarrow Fe^{3+}$$

   Therefore, the equivalent of ferrous iron equals 1 mol:
2. On the other hand, in the oxidation of metallic iron to ferric iron, the change in oxidation state is from 0 to 3

   $$Fe \rightarrow Fe^{3+}$$

   The equivalent of metallic iron, therefore, equals $\frac{1}{3}$ mol.
3. In volumetric analysis, when permanganate is used as an oxidizing agent in acid medium, permanganate ion is reduced into manganous ions:

   $$MnO_4^- \rightarrow Mn^{2+}$$

   or

   $$Mn^{7+} \rightarrow Mn^{2+}$$

   The change in oxidation state of Mn is from 7 to 2, which means a change in five units. The equivalent weight of permanganate is, therefore, $\frac{1}{5}$ mol. If used in neutral medium, the permanganate is reduced to MnO_2:

$$MnO_4^- \rightarrow MnO_2$$

or

$$Mn^{7+} \rightarrow Mn^{4+}$$

The state of oxidation of permanganate changes with three units. The equivalent then is $\frac{1}{3}$ mol.

As indicated in these examples, it is often not necessary to write the entire balanced equation of the reaction to find the equivalent weight. It is sufficient to write down only the change in oxidation state.

Another way to determine the equivalent weight (Eqw) follows. Instead of using the state of oxidation, the equivalent weight can also be found by the number of electrons transferred in the oxidation–reduction reactions. The following reactions serve as examples:

$$\text{Equivalent Weight} = \frac{\text{Molecular Weight}}{\text{Number of Electrons Lost or Gained}}$$

$$Fe^{2+} \rightarrow Fe^{3+} + e^- \qquad Eqw = \frac{Fe}{1}$$

$$Sn^{2+} \rightarrow Sn^{4+} + 2e^- \qquad Eqw = \frac{Sn}{2}$$

$$Fe(CN)_6^{4-} \rightarrow Fe(CN)_6^{3-} + e^- \qquad Eqw = \frac{Fe(CN)_6}{1}$$

$$As^{3+} \rightarrow As^{5+} + 2e^- \qquad Eqw = \frac{As}{2}$$

$$MnO_4^- + 8H^+ + 5e^- \rightarrow Mn^{2+} + 4H_2O \qquad Eqw = \frac{MnO_4}{5}$$

$$MnO_4^- + 4H^+ + 3e^- \rightarrow MnO_2 + 2H_2O \qquad Eqw = \frac{MnO_4}{3}$$

$$Cr_2O_7 + 14H^+ + 6e^- \rightarrow 2Cr^{3+}\ 7H_2O \qquad Eqw = \frac{Cr_2O_7}{6}$$

$$VO_4 + 6H^+ + e^- \rightarrow VO^{2+} + 3H_2O \qquad Eqw = \frac{VO_4}{1}$$

1.9 CHEMICAL UNITS

Normality

The number of a substance dissolved in 1 liter (L) of a solution determines the *normality* (N) of the solution. If one equivalent is present in 1 L of solution, the solution is 1 normal. The symbol N (normality) is usually italicized, or underlined, to distinguish it from the symbol N, the nitrogen element.

Molarity, Molality, and Formality

With the introduction of SI units, the use of concentration units expressed in terms of molarity, molality, or formality is preferred. The latter does not involve the determination of equivalent weights as with normality, N, hence, does not take into consideration the change in valencies or charges of the chemical compounds during a reaction process.

Molarity (M) is defined as the number of moles of a substance per liter of solution. *Molality* (m), on the other hand, is the number of moles per kilogram of water, and *formality* (f) is the number of moles of a substance per kilogram of solution. The symbols M and m are italicized, or underlined, to distinguish them from the SI unit terms M = mega and m = meter.

Molarity is expected to change in value with changes in temperature (T) and pressure (P), whereas molality and formality are independent of T and P. However, for very dilute solutions: $M = m$ = f, because the differences between them are then very small.

Mole Fraction

Another chemical unit, which is sometimes used, is the *mole fraction*. The latter is defined as the ratio of the number of moles of a given substance to the total number of moles of all constituents in solution, including the number of moles of water if it is an aqueous solution:

$$\text{Mole Fraction} = \frac{\text{Moles Ion}_1}{\text{Moles Ion}_1 + \text{Moles Ion}_2 + \cdots}$$

The following example serves as an illustration in the calculation of the mole fraction of NaCl in a 1 M solution:

$$\text{Mole Fraction NaCl} = \frac{1}{1 + 55.51} = 0.01769$$

In the foregoing equation, 55.51 is the moles of water in 1 L of water.

1.10   ISOTOPES

Isotopes are defined as the same elements with a similar atomic number, but with a differing mass number. It became apparent that not all atoms of the same element had the same atomic weight. Mass spectrographic analysis yielded evidence that oxygen could be separated into three types of oxygen with mass numbers of 16, 17, and 18, respectively. These types of oxygen are called isotopes of the oxygen element. They are different forms of the same element. The atoms of the isotopes have the same number of protons, but a different number of neutrons. Oxygen has three isotopes: ^{16}O, ^{17}O, and ^{18}O. However, 99.76% of all oxygen is in the form of ^{16}O. Hydrogen is known also to have three isotopes, $^{1}_{1}H$, $^{2}_{1}H$ (called *deuterium*), and $^{3}_{1}H$ (called *tritium*). Deuterium and tritium are rare. Water in which the hydrogen has been replaced by deuterium is called *heavy water*. Its formula is D_2O, and it has a density greater than H_2O (density = 1.076 g/mL at 20°C).

1.11   RADIOACTIVITY

Radioactivity, discovered for the first time in 1896 by Bacquerel, involves a spontaneous disintegration of certain types of elements or atoms to form other elements or atoms. Rutherford indicated that if atoms were able to eject particles, then these atoms were transformed into atoms with lower atomic weights. Ejection of α-particles means that two neutrons and two protons are emitted as a single particle. The remaining part of the atom is a new element. Among the many types of radioactive elements, perhaps uranium is the best known. By emission of an α-particle, the uranium nucleus loses two positive charges, and its atomic number drops from 92 to 90. Since the ejection of an α-particle also means a loss of two neutrons, the mass number of the new element decreases from 238 to 234. This new element is called *thorium*. In turn, thorium may emit electrons (β-particles), which is the result of a neutron decaying into a proton and an electron. A new element is then formed, called *proactinum*, with an atomic number of 91.

The emission of either electrons or α-particles from the nucleus of a decaying atom is a natural process. Before radioactivity stops, or the nucleus becomes stable, $^{238}_{92}U$ will continuously exhibit radiation by emission of α particles. The final product is usually lead 206 ($^{206}_{82}Pb$).

1.12 HALF-LIFE OF RADIOACTIVE MATERIAL AND CARBON DATING

Half-life is a measure of the rate of decay of radioactive material. It is the length of time during which the material loses one-half of its radioactivity.

Radioactive elements emit a definite number of particles per second. This rate of decay is dependent on the amount of radioactive material, and the rate of emission will, therefore, decrease as the amount of decaying atoms becomes gradually smaller and smaller as emission of particles progresses. For many of the radioactive materials, the time for complete decay is practically infinitely long, and it is very difficult, if not impossible, to determine the life span of these radioactive materials. However, the half-life span is relatively easily determined. The half-life of a radioactive element is a characteristic feature that cannot be altered by physical or chemical means. Examples of characteristic half-lives of selected elements are shown below (Weast, 1972):

Element	Half-life	Element	Half-life
U^{238}	4.5×10^9 yr	Co^{57}	270 days
U^{234}	2.5×10^5 yr	Fe^{55}	2.6 yr
Ra^{226}	1620 yr	Mn^{53}	2.0×10^6 yr
Pb^{207}	Stable	Ca^{45}	165 days
Pb^{206}	Stable	K^{42}	12.4 hr
Ba^{131}	12 days	Cl^{36}	3.1×10^5 yr
Zn^{65}	243.6 days	S^{35}	88 days
Cu^{64}	12.9 hr	P^{32}	14 days
Si^{31}	2.62 hr	Na^{22}	2.6 yr
Mg^{28}	21 hr	C^{14}	5730 yr
Al^{26}	7.4×10^5 yr		

Since the rate of decay for a given amount of radioactive material is constant, and since the final product is a stable Pb isotope, the ratio Pb concentration/^{238}U concentration is considered a measure for the minimum age of the sample. The foregoing principle is applied in the analysis for determining the age of geologic and anthropologic material with ^{14}C. This analysis is called *carbon dating*. Carbon 14, ^{14}C, is a radioactive isotope that has been formed by cosmic rays from outer space hitting nitrogen atoms in the atmosphere. The latter is then transmuted into ^{14}C, which decays into ^{12}C. An equilibrium usually exists in nature between the rate of ^{14}C formation and its rate of decay into ^{12}C. During their growth, the organism in nature obtains their carbon from the atmosphere directly or indirectly, and both ^{12}C and ^{14}C are absorbed in proportion to the equilibrium concentration existing in the air. As soon as these organisms die, the uptake of ^{12}C and ^{14}C ceases, but radioactive decay of ^{14}C into ^{12}C continues. By measuring the amount of ^{14}C in the material, the length of time since death can be estimated. Fresh or living plant material contains more ^{14}C than decomposed or death residue. Since the half-life of ^{14}C is only 5370 years, carbon dating is limited in its determination of age to approximately 10,000–15,000 years. For older materials, other radioactive methods are available.

2

Electrochemical Cells and Chemical Potentials

2.1 ELECTROCHEMICAL CELLS AND ELECTRODE POTENTIAL

The soil system is the reservoir of most plant nutrients and also contains the active surfaces that determine the concentration of ions in the soil solution. Ion movement, accumulation, availability of elements and uptake by plants, changes in element oxidation and reduction state, and many other chemical reactions in soils are reactions that, to a certain extent, show some resemblance to those occurring in an electrochemical cell. In pure chemistry two types of electrochemical cells are distinguished: (1) galvanic or voltaic cells and (2) electrolytic cells. A *galvanic cell* consists of two electrodes and one or more solutions (two half-cells). It is capable of spontaneously converting chemical energy from the solutions into electrical energy and supplying this energy to an external source. The automobile battery is an example of this kind of chemical cell. When one of the chemical components responsible for the reaction is depleted, the cell is considered dead. In an *electrolytic cell*, electrical energy is supplied from an external source. Electrochemical changes are produced at the electrode–solution interfaces (Figure 2.1), and concentration changes are developed in the bulk of the system. If the external current is turned off, the system tends to produce current in the opposite direction. The lowest external electromotive force (emf) that must be applied to bring about the continuous separation of cations and anions (electrolysis) is called the *decomposition voltage* or the *back emf.* At the exact point where the galvanic emf is opposed by an equal applied emf, no current flows in either direction. In this static condition, the potential generated at the interface of an indicator electrode reflects the composition of the solution phase.

Now that we know what electrochemical cells are, let us assume that we have a pure Cu electrode dipping into a solution containing cupric ions. At the interface, Cu tries to dissolve from the metal, increasing the Cu concentration in the solution. Therefore, a potential difference develops on the surface of the Cu electrode, between the metal and the solution. This potential difference is formulated by the Nernst equation

$$E = \frac{RT}{nF} \ln \frac{K}{M^n} \tag{2.1}$$

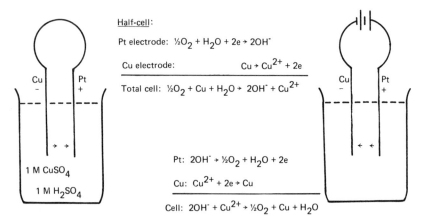

Figure 2.1 Schematic diagram of half-cell and total cell reactions in an electrolytic cell.

where

E = potential difference
R = gas constant
n = valence of the ion
F = Faraday constant
T = absolute temperature
K = impulse of metal to dissolve
M = ion activity

However, it is not possible to measure the potential of a single electrode. It is only possible to measure the potential of one electrode relative to another, the reference electrode. A standard reference electrode is the hydrogen electrode (Pt electrode dipping in a solution at unit activity of hydrogen ions). If the two electrodes are connected and an external emf is applied, electrons will flow through the system. Copper will be deposited on the Cu electrode:

$$Cu^{2+} (aq) + 2e^- \rightleftharpoons Cu (c)$$

At the hydrogen electrode, hydrogen will release electrons to become H^+:

$$H_2(g) \rightleftharpoons 2H^+ (aq) + 2e^-$$

Either of these reactions is called a *half-cell reaction*. The overall reaction is

$$Cu^{2+} (aq) + H_2(g) \rightleftharpoons Cu(c) + 2H^+ (aq)$$

The potential of the Cu electrode (half-cell) measured against the standard half-cell electrode is

$$E_h = E^0 + \frac{RT}{nF} \ln \frac{Cu^{2+}}{Cu} \tag{2.2}$$

At 25°C, this equation can be written as

$$E_h = E^0 + \frac{0.0592}{n} \log \frac{Cu^{2+}}{Cu}$$

E_h is called the *electrode potential*, whereas E^0 is the standard potential of the cell in which the reactants and products have unit activity. The electrode potential has been defined by the International Union of Pure and Applied Chemistry (IUPAC, 1960) as the electron availability of the electrochemical potential of the electron at equilibrium. The formula of E_h can be generalized as follows:

$$E_h = E^0 - \frac{RT}{nF} \ln \frac{\text{reduction}}{\text{oxidation}} \tag{2.3}$$

The standard potential of a cell when the other electrode is a standard hydrogen electrode is, by definition, called the *standard electrode potential*.

Depending on the way the reaction is written, electrode potentials are oxidation or reduction potentials. If the reaction is written as a reduction reaction,

$$Cu^{2+} \text{ (aq)} + 2e^- \rightleftharpoons Cu(c)$$

the standard potential is called a *reduction potential*. Both electrode and reduction potentials of Cu are then positive in sign. If, however, the reaction is written as an oxidation reaction:

$$Cu_{(c)} \rightleftharpoons Cu^{2+} \text{ (aq)} + 2e^-$$

the standard potential equals an oxidation potential and is negative in sign. The standard potential of the hydrogen electrode is by convention zero. Several selected potentials are given in Table 2.1 The difference of two standard potentials gives the standard potential of the desired reaction.

Electron availability, or redox potential, is an indication of the oxidation–reduction status of soils. It affects the oxidation states of H, C, N, O, S, Fe, Mn, Cu, and many other elements and, as such, controls solubility and availability of many nutrient elements to plants. But for the oxidation–reduction limits imposed by the stability of water, this list would include all the elements of the periodic table. The limit of oxidizing conditions in aqueous systems is the oxidation of water to molecular oxygen. On the other hand, the limit of reducing conditions is the reduction of the hydrogen ion to molecular hydrogen. In natural systems, redox potentials have often been treated as equilibrium potentials. At equilibrium, a mixture of redox couples reacts until the net donation and acceptance of

Table 2.1 Selected Electrode Potentials of Half-Cell Reactions

Half-cell reaction[a]	Electrode potential (V)
$3N_2 + 2H^+ + 2e^- \rightleftharpoons 2NH_3$	-3.100
$Li^+ + e^- \rightleftharpoons Li$	-3.045
$K^+ + e^- \rightleftharpoons K$	-2.924
$Ba^{2+} + 2e^- \rightleftharpoons Ba$	-2.900
$Na^+ + e^- \rightleftharpoons Na$	-2.710
$Mg^{2+} + 2e^- \rightleftharpoons Mg$	-2.375
$ZnO^{2-} + 2H_2O + 2e^- \rightleftharpoons Zn + 4OH^-$	-1.216
$Mn^{2+} + 2e^- \rightleftharpoons Mn$	-1.029
$Fe^{2+} + 2e^- \rightleftharpoons Fe$	-0.409
$Fe^{3+} + 3e^- \rightleftharpoons Fe$	-0.036
$2H^+ + 2e^- \rightleftharpoons H_2$	0.000
$Cu^{2+} + e^- \rightleftharpoons Cu^+$	$+0.158$
$AgCl + e^- \rightleftharpoons Ag + Cl^-$	$+0.222$
(silver–silver chloride electrode)	
Calomel electrode, $1N$ KCl	$+0.280$
$Cu^{2+} + 2e^- \rightleftharpoons Cu$	$+0.340$
$Fe^{3+} + e^- \rightleftharpoons Fe^{2+}$	$+0.770$
$Hg_2^{2+} + 2e^- \rightleftharpoons 2Hg$	$+0.789$
$Ag^+ + e^- \rightleftharpoons Ag$	$+0.799$
$MnO_2 + 4H^+ + 2e^- \rightleftharpoons Mn^{2+} + 2H_2O$	$+1.208$
$O_2 + 4H^+ + 4e^- \rightleftharpoons 2H_2O$	$+1.229$
$Cr_2O_7^{2-} + 14H^+ + 6e^- \rightleftharpoons 2Cr^{3+} + 7H_2O$	$+1.330$
$MnO_4^- + 8H^+ + 5e^- \rightleftharpoons Mn^{2+} + 4H_2O$	$+1.510$
$H_2O_2 + 2H^+ + 2e^- \rightleftharpoons 2H_2O$	$+1.770$

[a] In accordance to IUPAC conventions, the reactions above are written as reduction reactions. The signs of electrode potentials may perhaps be opposite in other books if written as oxidation reactions.
Source: Weast, 1972.

electrons is zero. The electron's escaping tendency or the E_h of each redox couple is the same, and their reduction/oxidation ratios have adjusted to that defined by the Nernst equation.

However, natural soil systems rarely reach oxidation–reduction equilibrium, because of the continuous addition of electron donors (i.e., oxidizable organic compounds). The oxidation of these compounds is often very slow, even when the major electron acceptor, oxygen, is available. Redox conditions in living systems are non-homogeneous, even within a single cell. In photosynthesis, for example, water is split into a strongly reducing form of hydrogen and the strong-oxidizing agent oxygen, within a chloroplast.

2.2 ELECTRON ACTIVITY

The derivation of formulas for the redox potential, as discussed in the preceding section, does not include the electrons. As noticed from the reactions, the electrons are vital participants in the reaction process. The problem is that free electrons, as such, do not exist in the soil solution, but neither do hydrogen ions. Both electrons and protons can exist only in close association with the solvent or a solute species. The analogy between electrons and protons can further be illustrated as follows: The activity of H^+ ions is used to indicate the acid–base condition of soil, as expressed in terms of pH. The soil is called acidic if the H^+ ion activity is high, in other words $H^+ >$ OH^- concentration. The soil is considered to be basic in reaction if the H^+ ion activity is low, or $H^+ < OH^-$ concentration. By analogy, then, soil with a high electron activity is considered to be in a reduced condition, whereas soil with a low electron activity is in a highly oxidized state. This can be further explained by using a generalized redox equation as follows:

$$\text{Oxidation} + e^- \rightleftharpoons \text{Reduction} \tag{2.4}$$

If electron activity increases, reduction increases (reaction shifts to the right). When, on the other hand, electron activity decreases, oxidation processes prevail (reaction shifts to the left). Application of the concept of electrode potential to the foregoing reaction yields an equation similar to Eq. (2.3):

$$E_h = E^0 + \frac{RT}{nF} \ln \frac{\text{oxidation}}{\text{reduction}} \tag{2.5}$$

or

$$E_h = E^0 + \frac{0.059}{n} \log \frac{\text{oxidation}}{\text{reduction}} \tag{2.6}$$

Many times, electron transfer in redox reaction is accompanied by a proton transfer. The following reactions serve as examples:

$$Al + 3H_2O \rightarrow Al(OH)_3 + 3H^+ + 3e^-$$

$$Fe(H_2O)_6 \rightarrow Fe(OH)_3 + 3H_2O + 3H^+ + e^-$$

From these reactions, it is now evident that electrons can also be considered as the reactants or products in a reaction process. Since H^+ activity can be expressed in terms of pH, hence, electron activity can also be represented by pe. In a way similar to how pH is defined: $pH = -\log H^+$, pe can be defined as:

$$pe = -\log (a_e) \tag{2.7}$$

In Eq. (2.7), a_e equals the activity of the electrons.

Some books use the symbols pE (Manahan, 1975), whereas other books prefer the use of P_e (Paul and Clark, 1989) for the negative log of electron activity. However, pE can refer to the negative log of the redox potential, E or E_h, whereas the symbol P ordinarily is used for pressure or potential. It is an established fact, in chemistry, thermodynamics, and the pH concept, that the symbol p is meant to indicate $-\log$. Therefore, to avoid confusion and to preserve consistency, the symbol pe is preferred in this book over the use of pE or P_e.

The application of the pe concept is based on the reaction

$$2H^+_{(aq)} + 2e^- \leftrightharpoons H_{2(g)} \tag{2.8}$$

This reaction is used for defining the free energy changes, ΔG, in redox processes in soil. When all the components of the foregoing reaction are at unity, ΔG for this reaction equals zero. Therefore, when both the H^+ ion and the H_2 gas are at unit activity, the activity of the electron is also 1 (or unity); hence, the $pe = -\log 1 = 0$. The electrode potential for Eq. (2.8) then also equals zero (see Table 2.1). If the electron activity were to increase to 10, the pe would become -1.0. If the electron activity were to decrease to -10, the pe would become $+1.0$.

2.3 RELATIONSHIP BETWEEN ELECTRON ACTIVITY AND ELECTRODE POTENTIAL

As will be shown in Chapter 3, the electrode potential E_h is related to the negative log of the electron activity, pe, by the equation:

$$E_h = \frac{2.3RT}{F} \text{ pe} \tag{2.9}$$

where R = the gas constant, T = absolute temperature (Kelvin; K) and F = Faraday constant. Since these constants are known, at T = 298 K (or 25°C), the equation may be changed into:

$$E_h = 0.059 \text{ } pe \tag{2.10}$$

or

$$pe = E_h/0.059 \tag{2.11}$$

In this equation E_h is in volts (V). A similar relationship can be found for the standard electrode potential: $E^0 = 0.059 \text{ } pe^0$.

The electrode potential can, therefore, be expressed also in terms of pe, and conversion of E_h into pe (or E^0 into pe^0) can be easily accomplished by using Eqs. 2.10 and 2.11. A partial list of half-cell reactions and their standard electrode potentials from Table 2.1 is presented as follows to illustrate the conversion of E^0 into pe^0:

Half-cell reaction	E°	pe°
$Mn^{2+} + 2e^- \leftrightarrows Mn$	-1.029 V	-17.44
$Fe^{2+} + 2e^- \leftrightarrows Fe$	-0.409 V	-6.93
$2H^+ + 2e^- \rightleftarrows H_2$	0.000 V	0.00
$Cu^{2+} + 2e^- \leftrightarrows Cu$	$+0.340$V	5.76
$Fe^{3+} + e^- \leftrightarrows Fe^{2+}$	$+0.770$ V	13.05

The more positive the value of E^0 or pe^0, the greater the tendency for these reactions to proceed to the right.

2.4 CHEMICAL POTENTIALS AND THEIR APPLICATION IN ION UPTAKE BY PLANTS

Chemical Potential

To each chemical species in a reaction mixture, a certain amount of (free) energy can be ascribed. This amount of energy, expressed per unit amount of ion species, is called *chemical potential.* This entity, indicated by the symbol μ, depends on the pressure P, the temperature T, the chemical nature of the ion species, and on its mixing ratio with other species. For ideal solutions, the chemical potential can be formulated as follows:

$$\mu = \mu^0 + RT \ln m \qquad (2.12)$$

where

μ = chemical potential of an ion species
μ^0 = chemical potential of the ion species at standard state
m = moles of ion species in the mixture
R = gas constant
T = absolute temperature (K)

In some books, the chemical potential is defined as:

$$\mu = \mu^0 + RT \ln m/m^0 \qquad (2.13)$$

In this equation m^0 equals m at standard state, which is considered a hypothetical molality in which all interactions leading to deviations from ideal conditions have been canceled. Consequently, m^0 = 1, and Eq. 2.13 converts automatically into Eq. (2.12). For non-ideal conditions, Eq. (2.12) changes into:

$$\mu = \mu^0 + RT \ln a \qquad (2.14)$$

in which a is the activity of ion species.

At infinite dilution (or very dilute) conditions, Eq. (2.14) can be changed into:

$$\mu = \mu^0 + RT \ln c \qquad (2.15)$$

In this equation, c denotes the concentration of the ion species. Since μ^0 is considered a constant, the chemical potential can be written as $\mu = RT \ln m$ (ideal), $\mu = RT \ln a$ (nonideal), and $\mu = RT \ln c$ (very dilute condition) plus a constant.

For practical purposes, the equations for the foregoing chemical potentials can be converted into "$\log_{10}$" form, and Eq. (2.15), for example, can be changed into:

$$\mu = \mu^0 + 2.3RT \log c \qquad (2.16)$$

Since R is a known constant, at $T = 25°C = 298$ K, Eq. (2.16) converts automatically into:

$$\mu = \mu^0 + 1.364 \log c \tag{2.17}$$

or

$$\mu = 1.364 \log c + \text{constant} \tag{2.18}$$

Equilibrium Potential

As stated earlier, the chemical potential indicates the state of potential energy of the chemical species or component in soils, and its formulation shows some relationship with the Nernst potentials. It is independent of external force fields, such as gravity or centrifugal forces, and the chemical potential will remain constant at any distance from the soil surface. It also remains constant at equilibrium conditions (e.g., constant concentrations, temperature, and pressure). Differences may occur at nonequilibrium conditions (e.g., owing to differences in concentrations). These differences in chemical potentials of a species at various locations in soils tend to induce spontaneous movement of the species in the direction of points with lower potentials until an equilibrium condition is attained.

Application of the concept of chemical potentials in a hypothetical soil condition shows the following relationships: In soils, ions are adsorbed by clays and humic acids. Small molecules (ions) are then bonded by large molecules (clay and humus). If such a system is placed in a dialysis bag (inside $= i$), H_2O and the ions will move freely through the semipermeable membrane of the dialysis bag into the solution outside (o), but not the clay and humic acid molecules. The movement of the ions continues until an equilibrium condition is attained, at which the differences in chemical potentials "inside" and "outside" are erased. Consequently, at equilibrium, the chemical potential of the ions inside the bag equals the chemical potential of the ions outside the bag. Ignoring the presence of electrical potentials the following relationship is valid:

$$\mu_i = \mu_o \tag{2.19}$$
$$RT \ln a_i = RT \ln a_o$$
$$a_i = a_o \tag{2.20}$$

Since $a = \gamma c$, as discussed in Chapter 3, activity can be changed into concentration units, and Eq. (2.20) becomes:

$$\gamma_i c_i = \gamma_o c_o \tag{2.21}$$

Membrane or Donnan Potential

The chemical potential, as formulated in the preceding section, can also be applied to describing ion transport, movement, and adsorption in soil. In addition, it is particularly useful in predicting ion uptake by plant roots and ion transport from cell to cell in the plant body. The selective uptake of chemical compounds by the root system is an outstanding process of living systems. Cell compartments in the plant tissue are separated by biological membranes representing barriers for chemical compounds. The transport through these barriers and processes mediating this transport can be studied and described using chemical potentials. Although not clearly understood, it is currently accepted that plant cells behaving as biological membranes have pores acting as sieves, favoring penetration of small particles only. This kind of passive transport (diffusion, mass flow) of small hydrophilic particles obeys physical and chemical laws. The net ion flux in either direction will stop as soon as a state of equilibrium is reached. At equilibrium, the system conforms to the Donnan principle. With compounds that are not electrically charged (e.g., sucrose), the equilibrium is attained when equal sucrose concentrations exist on either side of the membrane. However, with ions possessing electrical charges, the electrochemical potentials must be equal on both sides of the membrane:

$$\mu_i + zF\psi_i = \mu_o + zF\psi_o \tag{2.22}$$

where
$\quad i$ = inside cell
$\quad o$ = outside cell
$\quad \psi_i$ = electrical potential at innerside of membrane
$\quad \psi_o$ = electrical potential at outerside of membrane
$\quad \mu$ = chemical potential of element species
$\quad z$ = valence
$\quad F$ = Faraday constant

The chemical potential was defined earlier as $\mu = RT \ln a$; therefore, substituting $RT \ln a$ for μ, Eq. (2.22) becomes

$$(RT \ln a)_i + zF\psi_i = (RT \ln a)_o + zF\psi_o$$
$$zF\psi_o - zF\psi_i = (RT \ln a)_i - (RT \ln z)_o \qquad (2.23)$$
$$\psi_o - \psi_i = E = \frac{RT}{zF} \ln \frac{a_i}{a_o}$$

$\psi_o - \psi_i$ is called the *membrane* (or *Donnan*) *potential*, and by chang-ing it into symbol E, Eq. (2.23) conforms to the Nernst equation. Because $\psi_o - \psi_i$ is a positive figure, we have

$a_i > a_o$ for cations

$a_i < a_o$ for anions, z is negative

At equilibrium, the chemical potential of cations also equals the chemical potential of anions. This is necessary to maintain eletro-neutrality. Therefore,

$$E_c = \frac{RT}{zF} \ln \left(\frac{a_i}{a_o}\right)_c \text{ equals } E_{an} = \frac{RT}{zF} \ln \left(\frac{a_o}{a_i}\right)_{an}$$

where

c = cations
an = anions

Consequently,

$$\left(\frac{a_i}{a_o}\right)_c = \left(\frac{a_o}{a_i}\right)_{an} = \text{constant} \qquad (2.24)$$

The latter means that the ion product on either side of the membrane is constant. If KCl or $CaCl_2$ is present in the system, then according to the Donnan equilibrium principle the following is valid:

$$(K^+)(Cl^-)_{inside} = (K^+)(Cl^-)_{outside}$$

or

$$(\sqrt{Ca^{2+}})(Cl^-)_{inside} = (\sqrt{Ca^{2+}})(Cl^-)_{outside}$$

Once again, one needs to remember that the foregoing hypothesis has been developed for equilibrium condition. One must also take into consideration that in most metabolically active living cells, no equilibrium will exist between the two sides of the membrane. Metabolism is continuously consuming ions on the inside and, thereby, it is constantly disrupting the equilibrium condition favoring passage of ions from the outside.

3

Soil Composition and Soil Solution

3.1 SOIL COMPOSITION

The soil system is composed of three phases: solid, liquid, and gas. The solid phase is a mixture of mineral and organic material and provides the skeletal framework of the soil. Enclosed within this framework is a system of pores, shared jointly by the liquid and gaseous phase. The spatial arrangement of the solid particles and associated pores, or voids (Figure 3.1) is, in micropedological terms, the *soil fabric* (Kubiena, 1938; Brewer and Sleeman, 1960). The nature of the soil constituents, the activity of soil organisms, and other soil-forming processes leads to the creation of a series of features with specific soil fabrics. Although the traditional concept indicates that soil fabric is comparable with rock fabric, which depends on the shape and arrangement of minerals, it is perhaps more appropriate to compare soil fabric with the fabric of plant and animal tissue. The components of the soil fabric include water and organic substances, containing humus and organisms, and a series of organic and inorganic compounds. These materials are also of importance in plant and animal tissue. Rocks do not contain these components.

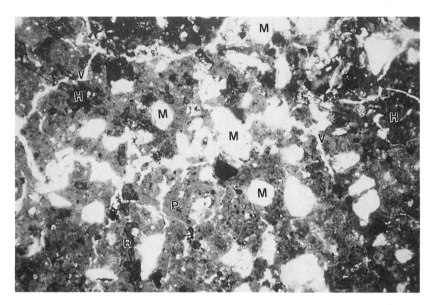

Figure 3.1 A soil thin section, showing the spatial organization of soil constituents and associated pores (voids), called *soil fabric*. The pores contain water and air. M, primary minerals; H, humus; P, plasma (clay); R, root fragments or vegetative remain; and V, void or pore. (Magnification × 40.)

3.2 SOIL AIR

Depending on the conditions, the total pore space may vary from approximately 25% (Manahan, 1975) to 50% by volume in loam surface soils optimum for plant growth (Brady, 1990). As indicated before, the pores are filled with air and water. The composition and chemical behavior of this liquid and gas phase are determined by the interaction with the solid phase. The gaseous phase, or *soil air* is a mixture of gases. The content and composition of soil air is determined by soil–water–plant relationships. Whereas atmospheric air contains 20% oxygen, 0.3% carbon dioxide, and 78% nitrogen by volume, soil air tends to be lower in oxygen and higher in carbon dioxide content. Two biological reactions are responsible for this condition in soil:

1. *Respiration* of plant roots and soil organisms: Plant roots and soil organisms absorb oxygen and give off carbon dioxide. This exchange of gases, called respiration (Martin et al, 1976), involves the breakdown of sugars by oxidation. The process can be represented by the reaction:

$$C_6H_{12}O_6 + 6O_2 \rightarrow 6CO_2 + 6H_2O + \text{energy} \qquad (3.1)$$

The energy released is needed by the roots and microorganisms for performing work (metabolism) and growth.

2. *Aerobic decomposition* of organic matter: Decomposition is an essential reaction in the carbon cycle. The reaction also consumes O_2 and produces CO_2. In this process, soil organic matter is broken down into CO_2 and H_2O by an oxidation reaction, similar to Eq. (3.1). The energy, released in the form of heat, is usually noticed by the increase in temperature underneath a composting pile of organic matter.

Unfortunately no biochemical reactions are present in the soil that can replenish the depleted O_2 content in soil air or correct the O_2 and CO_2 balance. Photosynthesis is the only biological reaction known to produce oxygen and to consume carbon dioxide. However, photosynthesis occurs in plant parts growing above the soil. The green plants are capable of absorbing excess CO_2 from pollution and releasing O_2 in the atmosphere. Hence, they act as a cleaner for polluted air, and their destruction, as occurs by large-scale deforestation, for example, may decrease the quality of air that we breath. Therefore, the O_2 and CO_2 content in soil air can be corrected only by exchange of soil air for atmospheric air, a process called *aeration*.

3.3. CONCEPT AND IMPORTANCE OF SOIL SOLUTION

The liquid phase, also designated as the *soil solution*, is composed of water, colloidal material, and dissolved substances. The substances are free salts, and often the ions of these salts are attached to clays, to other colloidal material, to organic solutes, or to a combination thereof. Depending on the forces acting on soil water, the water may be free to move, but the movement of solutes may be more or less constrained or may also effect some constraint on the movement of water. Soil water also contains several gases, although the solubility of these gases is relatively small. *Henry's rule* indicates that the solubility of a gas in a liquid is proportional to the partial pressure of the gas in contact with the liquid. Water vapor itself possesses a partial pressure that influences gas solubility. Two important types of gases, which may dissolve in soil water, are oxygen and carbon dioxide. Whereas the oxygen content in soil water is vital for the growth of many organisms, such as aerobic bacteria, carbon dioxide may react with water to form H_2CO_3. As a weak acid, H_2CO_3 may affect the soil pH. The soil solution as described is the medium in which most soil chemical and biochemical reactions occur. It bathes the plant roots and forms the source from which the roots and other organisms obtain their inorganic nutrients and water. Therefore, the soil solution provides the chemical environment of plant roots, and defining plant–soil–water interrelations in quantitative terms requires a complete and accurate knowledge of soil solution chemistry and the laws governing it.

3.4   CHEMISTRY OF SOIL WATER

Water is a renewable resource that belongs to a gigantic cycle, the *hydrologic cycle*. It is evaporated from the earth and the sea into the atmosphere. The energy required for evaporation is supplied by the sun. In the atmosphere, water is transported by the wind in the form of vapor, until it finally returns to the earth in some form of precipitation.

Until late in the 19th century, water was thought to be an element. It was Cavendish in 1871 who showed that water could be formed by burning H_2 in the air. A few years later Lavoisier determined the composition of water as H_2O.

In its common state, water is a colorless, odorless, and tasteless liquid. It may also exist in a vapor or solid state, and often all three phases can occur at the same time. Some of the properties are peculiar to its liquid state, whereas other properties may be more common to the vapor and solid states. A single water molecule is very small, generally in the dimension of 3 Å (3×10^{-8} cm or 0.3 nm) in diameter. The unit mass of H_2O is the mass of 1 mL of H_2O, which equals 1 g (4°C). One mole of water (18 mL) contains 6.02×10^{23} individual molecules. Water participates directly in many soil and plant reactions and, indirectly, affects many others. Its ability to react is determined by its chemical structure. An individual water molecule is composed of one atom of oxygen attached to two atoms of hydrogen. The hydrogen atoms are at an angle of about 105° from each other (Figure 3.2). This arrangement causes an imbalance of charges, with the center of the positive charge being at one end and the center of the negative charge at the other end. Such molecules are called *dipolar* because of their behavior in electrical fields, according to the imbalanced charges.

When water crystallizes, the molecules arrange themselves so that a hydrogen atom of one water molecule is located close to an oxygen atom of another molecule of water. The bond by which a hydrogen atom acts as the connecting linkage is called a *hydrogen bond*. As a result of crystallization, water forms a hexagonal structure with many empty spaces. Consequently, ice is less dense than liquid water and will float in water. The formation of empty spaces causes a volume expansion of about 9%, and the forces developed by the latter is approximately 150 kg/cm^2. Therefore, when water freezes in soils and plants, the expansion causes the soil structure to change and the plant cells to rupture.

The presence of dipoles in the water molecules accounts for a number of other important reactions. Cations (e.g., Na^+, K^+, Ca^{2+}) become hydrated through their attraction to the negative pole of water molecules. Likewise, negatively charged clay particles attract water through the positive pole of the water molecule. Polarity of water also encourages dissolution of salts, as the ionic components of salts have greater affinity for water molecules than for each other. When water molecules become attracted to clay surfaces or ions,

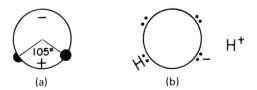

(a)   (b)

Figure 3.2 (a) A water molecule composed of 2H atoms attached at 105° angle to an O atom; and (b) ionization of water into a hydroxyl (OH^-) and a proton (H^+) or hydrogen ion. Dots are unshared electron pairs, except where H is linked to O.

they do so in packed clusters. In clusters, the free energy of water is lower than in "free" water, meaning the molecules are less free to move. Free water has greater internal energy than ice or clusters of water. The latter is indicated by the release of 80 cal (334 J) of energy in the form of heat as water freezes into ice. The SI unit of heat is the quantity of heat required to raise the temperature of 1 g of water from 14.5° to 15.5°C (1 calorie = 4.18 Joules). In scientific terms, the energy differences mean that the entropy of water is lower in the solid than in the liquid state. When ions become hydrated, energy is released, a phenomenon called *heat of solution*. When clay particles are hydrated, the energy released is called *heat of wetting*. Surface tension is another important property of water that influences its behavior in soils. This property occurs only at the liquid–air interfaces and is the result of a greater attraction between water molecules for each other than for the air above. The net effect is an inward force of water molecules at the surface, causing water to behave as if its surface were covered with an elastic membrane. Surface tension plays an important role in soil–water movement, called *capillarity*.

3.5 OXYGEN DEMAND OF WATER

As indicated earlier, oxygen dissolves in water and is an important substance for the growth of many organisms. It is often considered the key substance for the existence of many kinds of life in water. Oxygen deficiency is fatal to fish and to many types of aerobic bacteria.

The solubility of oxygen in water is affected by the water temperature, the partial pressure of oxygen in the atmosphere (or in soil air), and the salt content of water. The amount dissolved is, therefore, very small, and water at 25°C and 1 atmosphere pressure (atm) contains only 8.32 mg dissolved O_2 per liter (Manahan, 1975). This amount of dissolved oxygen may be depleted very rapidly unless some mechanism for aeration is present. The latter can be achieved naturally by turbulent flow in streams or, artificially, by pumping air in water such as in aquariums and sewage treatment plants. Several processes are responsible for the depletion of dissolved oxygen content. Oxygen can be consumed rapidly in the oxidation process of organic matter as illustrated by Eq. (3.1). Other microor-

ganism-mediated processes depleting the dissolved oxygen content
are the *biochemical oxidation* of

1. Iron compounds:

$$4Fe^{2+} + O_2 + 4H^+ \rightarrow 4Fe^{3+} + 2H_2O \qquad (3.2)$$

2. Sulfur compounds:

$$2SO_3^{2-} + O_2 \rightarrow 2SO_4^{2-} \qquad (3.3)$$

3. Ammonium:

$$2NH_4^+ + 3O_2 \rightarrow 2NO_2^- + 2H_2O + 4H^+ + energy \qquad (3.4)$$

$$2NO_2^- + O_2 \rightarrow 2NO_3^- + energy \qquad (3.5)$$

The oxidation of ammonium into nitrate, which occurs in two steps,
is called *nitrification*.

The rate by which oxygen is consumed or depleted by the foregoing processes is called *oxygen demand* (OD). Three types of ODs
are distinguished.

1. *Biological oxygen demand (BOD).* This is defined as the
   amount of oxygen consumed by microorganisms in the process of decomposition of organic matter during a 5-day in-
   cubation period. The test was developed in England, where
   it was believed that dissolved organic matter not decomposed
   within 5 days would be transported into the sea.
2. *Chemical oxygen demand (COD).* This is defined as the
   amount of oxygen consumed in the chemical oxidation of organic matter by potassium dichromate in the presence of sulfuric acid. The analysis for oxidation of organic C itself is
   known today as the Walkley–Black method.
3. *Total oxygen demand (TOD).* This is the amount of oxygen
   consumed in the catalytic oxidation of carbon. The amount
   of CO_2 produced in the reaction is measured.

Of these three types of ODs, the BOD is the best known, although
its determination is more difficult than the determination of COD,
whereas the 5-day period is subject to many arguments. Neverthe-
less, many scientists prefer the use of BOD for determining the quality of stream and lake water or the amount of pollution in soils and
the environment. The lower the BOD values, the better will be the

water quality. A BOD value of 1 ppm (which means 1 ppm of oxygen consumed in the decomposition during a 5-day period) indicates that presence of low amounts of oxidizable organic contaminants, in other words: water of high quality. On the other hand, high BOD values (5–20 ppm) indicate the presence of water with high amounts of organic contaminants, or water with low quality (Stevenson, 1986; Manahan, 1975). In view of the maximum amount of oxygen that can be dissolved in water (8.32 mg/L, at 25°C and 1 atm), a value of 8 ppm as the lower limit of a high BOD value is perhaps closer to reality than the value of 5 ppm.

3.6   SOIL–WATER ENERGY CONCEPTS

Water in soils may possess different kinds and quantities of energy. Differences in energy content of water at various locations in soils cause it to flow. Retention of water in soils, its uptake and transport in plants, and the loss of water to the atmosphere, all are processes that are related to changes in the energy status of soil water. Several kinds of energy are usually involved: potential, kinetic, and electrical energy (Brady, 1974). However, Hillel (1972) indicated that only potential and kinetic energy were traditionally recognized as the two principal forms of energy in physical science. Of the two, potential energy is considered of more importance in determining the state and movement of water in soils. Differences in potential energy from one to another point in the soil tend to create movement of soil water. The water flow is in the direction of decreasing potential energy. The force causing this flow of water is called *water potential difference* (because of a difference between two points). In contrast, a water potential gradient is a continuously changing function of potential in a flow medium. The flow of water continues until the water potential difference between the two points is zero, and an equilibrium condition is attained.

Water Potential (ψ_w)

The term *water potential* was introduced by Buckingham in 1907, who used it as a synonym for capillary potential (Hillel, 1972). Currently, it is applied to describe water's energy status or ability to work. Work is considered performed as water moves from one to another location in soils. At equilibrium condition, water has the potential to do this work, and the energy associated with it is called *potential energy* or briefly *potential*. This water potential is the net effect of several components. It is the sum of the contributions of the various forces acting on soil water, such as matric, osmotic, and solute forces. For isothermal conditions, the water potential can be formulated as follows:

$$\psi_w = \psi_m + \psi_p + \psi_s \qquad (3.6)$$

where

ψ_m = matric potential
ψ_p = pressure potential
ψ_s = solute potential and soil water

This water potential can assume a positive or a negative value depending on the forces acting on soil water. The presence of solutes and matrix components decreases the capacity of water to perform work; hence, under normal field conditions, the soil water potential is negative. However, under conditions for which the hydrostatic pressure is greater than the atmospheric pressure (in pressure plates), the water potential will be positive.

From the previous discussion, it is apparent that the relative level of energy of water at different locations in soils is of more value to the behavior of soil water than the absolute amount of energy that soil water contains. As indicated earlier, the term *water potential* is used as a measure to express this energy level of soil water relative to that of water at standard rate. If external force fields are excluded, this concept of water potential shows considerable analogy with the concept of chemical potentials. As discussed before, differences in chemical potentials of a component between two points in the soil also govern the direction of movement of the component (here the movement of soil water). Not surprisingly, several authors have tried to apply thermodynamic concepts of chemical potentials in soil–water problems, although several arguments questioning its applicability were frequently reported (Taylor and Ashcroft, 1972; Hillel, 1972; Taylor and Slatyer, 1960). From thermodynamic principles, the water potential is considered to be equal to the difference between the chemical potential of soil water at an arbitrary equilibrium state and that of soil water at standard state (Taylor and

Ashcroft, 1972). That relationship can be expressed as follows:

$$\psi_w = \Delta\mu_w = \mu_w - \mu_w^0 \tag{3.7}$$

where

$\Delta\mu_w$ = difference in chemical potential of soil water
μ_w = chemical potential of soil water at arbitrary
        equilibrium condition
μ_w^0 = chemical potential of soil water at standard state

Formerly $\Delta\mu_w$ was called the *moisture potential* (Taylor and Slatyer, 1960).

Total Soil–Water Potential (ψ_t)

The *total soil–water potential* is defined by the International Soil Science Society, Soil Physics Terminology Committee (Aslyng et al., 1963) as the amount of work required to transport reversibly and isothermally an infinitesimal amount of water from a pool of pure water at a specified elevation at atmospheric pressure to the point under consideration. This potential includes the water potential and potentials arising from external force fields. It is generally formulated as follows:

$$\psi_t = \psi_w + \psi_g + \psi_z + \cdots$$

where

ψ_w = water potential
ψ_g = gravitational potential
ψ_z = any potential arising from external force fields

Since, in soils, gravity is the only external force field of importance, the total water potential ψ_t is usually expressed as

$$\psi_t = \psi_w + \psi_g \tag{3.8}$$

Depending on the forces acting on soil water, the total water potential can also assume positive or negative values.

Matric Potential (ψ_m)

The attraction of soil solids (matrix) for water provides a matrix force, and that part of the water potential attributed to the matrix force is called *matric potential*. Taylor and Ashcroft (1972) reported that the matric potential is similar to the capillary potential and replaces terms such as soil moisture tension, soil moisture suction, or matric suction. The matrix force reduces the free energy of the adsorbed water. In the presence of solid particles (matrix), water is subject to adsorption on the particle surfaces, and the adsorbed water cannot move as freely as free water. The matric potential can be determined with a tensiometer.

Pressure Potential (ψ_p)

Pressure differences in soils, resulting from air or pneumatic pressures of the atmosphere on soil water, are the reasons for the development of pressure potentials. In a saturated soil, the pressure potential has a positive value because the hydrostatic pressure is greater than the atmospheric pressure. In a water-unsaturated soil, the pressure potential equals zero, since the liquid (water) pressure can be neglected, whereas the soil air pressure can be considered equal to the atmosphere. A negative pressure potential occurs only when soil water is subjected to a pressure lower than atmospheric pressure. Taylor and Ashcroft (1972) stated that negative pressures were normally observed in laboratory conditions. However, Hillel (1972) indicated that the negative pressure potential was synonymous with capillary and matric potential, and was of the opinion that both capillary and adsorptive forces were responsible for the negative pressure potential. Thus, the matric potential, or the negative pressure potential, is a reflection of the total effect resulting from the retention of water in soil pores (capillary forces) and on the surfaces of soil particles (adsorption).

Osmotic Potential (ψ_o)

The portion of the water potential attributed to the attraction of solutes (ions and other molecules) for water by osmotic forces is called *osmotic potential*, sometimes also known as *solute potential*.

This potential becomes of importance only if a semipermeable membrane is present. The latter acts as a barrier for movement of the solutes, but water can flow freely through the membrane. In the absence of a membrane, the solutes will flow with the water, instead of the solutes attracting water.

Gravitational Potential (ψ_g)

Gravitational potential is the portion of the total water potential attributed to the downward pull of water by gravity. By moving water against the gravitational force, work is performed. The amount of work needed is stored in the water in the form of potential energy. The gravitational potential is independent of the chemical potential, but depends on the vertical location and the density of soil water.

Units of Soil-Water Potential

The units for soil-water potential can be expressed in several ways (Table 3.1). Usually the water potential is expressed in units of contained energy per unit mass of water (ergs per gram of water, or joules per kilograms of water). Sometimes, the water potential is stated in units of energy per mole of water (ergs per mole). The latter is also known as *molar water potential*. Another way to express soil

Table 3.1   Units of Soil-Water Potentials and Their Equivalents

Energy/unit mass		Volumetric potential,	Soil-water suction	Relative humidity
(erg/g)	(J/kg)	(bars)	(bars)	(%)
0	0	0	0	100.00
-1×10^4	-1	-0.01	0.01	100.00
-5×10^4	-5	-0.05	0.05	99.99
-1×10^5	-10	-0.10	0.10	99.99
-3×10^5	-30	-0.30	0.30	99.97
-5×10^5	-50	-0.50	0.50	99.96
-1×10^6	-100	-1.00	1.00	99.92
-5×10^6	-500	-5.00	5.00	99.63
-1×10^7	-1000	-10.00	10.00	

Source: Hillel (1972), Taylor and Ashcroft (1972), Weast (1972).

water potential is in units of energy per volume of water (ergs per cubic centimeter), which is then called *volumetric water potential*. The latter unit is equivalent to the unit of pressure in physics, dynes per square centimeter (1 erg = 1 dyn/cm^2).

3.7 PLANT–SOIL–WATER ENERGY RELATION

The theories of water potentials are well established and have proved useful in describing the conditions of water in plant tissue and the movement of water through the plant.

Water is retained in plant cells by adsorptive and osmotic forces. The cells consist of (1) cell walls, which are rigid, but exhibit a capability for elastic expansion; (2) protoplasm, acting as a semi-permeable membrane through which water can move freely, and in contrast with the flow of solutes, which are somewhat restricted; and (3) vacuoles, filled with solute-rich cell sap and some colloidal material. The solute and colloidal concentration reduce the activity of water inside the cell. The higher solute and colloid concentration results in attraction of water, and water outside the membrane will move into the cell more rapidly than the solutes can diffuse out. The attraction of the cell for water from solute concentration in the vacuole is called *solute potential* ψ_s. If the attraction arises from the adsorption of water by colloidal material in the cell, or by proto-plasmic colloids, it is called *matric potential* ψ_m. The combination of solute and matric potential is the *osmotic potential* ψ_o. The turgor pressure accompanying the adsorption of water by the cell is called *turgor*, or *pressure potential* ψ_p. This is the potential forcing water out of the cell as a result of inflated conditions of the cell. When water moves into a cell, the cell volume increases (Figure 3.3), and the protoplast is forced against the cell wall, which, being elastic, expands. The greater the expansion of the cell, the greater will be the pressure exerted on water within the cell, and the turgor pres-sure increases accordingly. Consequently, the flow of water into the cell decreases gradually as the turgor pressure ψ_p within the cell increases. At one point, the pressure (turgor) potential can become numerically equal, but opposite in sign, to the combined solute and matric potentials. At the point of equal pressure, the sum of poten-tials $\psi_p + \psi_s + \psi_m$ is then equal to zero, and the net flux of water

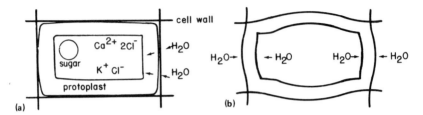

Figure 3.3 (a) Schematic diagram of plant cell showing vacuole filled with cell sap, sugar, Ca^{2+}, K^+, and Cl^- ions. Water moves freely into cell (from Taylor and Ashcroft, 1972); and (b) inflated cell with increased turgor potential exerting pressure on water within the cell. Movement of water into and out the cell stops when turgor equals osmotic potential.

into the cell becomes zero. Flux refers to the rate of flow. The sum of the solute, matric, and pressure (turgor) potentials is called the *water potential* ψ_w:

$$\psi_w = \psi_s + \psi_m + \psi_p \tag{3.9}$$

There is also a tendency for water to move through the cell membrane into the soil. This movement can be prevented by increasing the pressure or decreasing the tension on the soil side of the membrane.

3.8 THE LAW OF MASS ACTION AND THE EQUILIBRIUM CONSTANT

Almost all chemical and biochemical reactions occur in aqueous solutions. The processes will be governed by basic chemical laws, one of which is the law of mass action. Consider the reaction

$$A + B \rightleftharpoons C + D$$

in which A and B are the reactants, and C and D are the reaction products. The rate of reaction from left to right (R_1) is proportional to the product of the concentrations of A and B:

$$R_1 = k_1 C_A \times C_B$$

Similarly the rate of reaction from right to left (R_2) can be written as

$$R_2 = k_2 C_C \times C_D$$

in which k_1 and k_2 are proportionality constants.

At equilibrium R_1 must equal R_2:

$$R_1 = R_2$$

$$k_1 C_A \times C_B = k_2 C_C \times C_D \tag{3.10}$$

$$\frac{C_C \times C_D}{C_A \times C_B} = \frac{k_1}{k_2} = K_{eq}$$

K_{eq} is called the *equilibrium constant*. This equation is the formulation of the mass action law as first reported by Guldberg and Waage in 1865. A revised definition was, however, provided by van't Hoff in 1877, and since then this law has been applied to various systems. The *law of mass action* says that at equilibrium the product of the concentrations of the reaction products divided by the product of the concentrations of the reactants is constant in any chemical reaction. In a concentrated solution the concentrations in Eq. (3.10) must be replaced by activities. However, for a first approximation, activities may be replaced by concentrations in moles per liter. The equilibrium constant K_{eq} depends on the temperature of the solution, not on the pressure of the air or the composition of the system. It is fixed for any given temperature. The larger the numerical value of K_{eq}, the greater the tendency for the reaction to proceed in the direction of the reaction products.

The aforementioned law of equilibrium finds extensive application in many ionic solution phenomena, and the equilibrium constant assumes different kind of expressions (e.g., solubility product constant, dissociation constant or ion product constant of water, ionization constant of electrolytes, cation exchange constant, or ion exchange constants).

3.9 SOLUBILITY PRODUCT

The *solubility product* is the product of ion concentrations in the saturated solution of a difficulty soluble salt. Consider the dissociation of BA occurring as follows:

$$BA \rightleftharpoons B^+ + A^-$$

Since the activity of pure solids is unity at equilibrium, application of the mass action law gives

$$K_{eq} = K_{sp} = (B^+)(A^-) \qquad (3.11)$$

In this type of reaction K_{eq} is called the *solubility product constant* and the symbol K_{sp} is used. The negative log of K_{sp} is pK_{sp}:

$$pK_{sp} = -\log K_{sp}$$

The smaller the pK_{sp}, the more soluble the substance. However, if two solutions, each containing one of the ions of a difficulty soluble salt, are mixed, no precipitation will take place unless the product of the ion concentration in the mixture is greater than the solubility product.

In saturated solution the concentration of B^+ ions is equal to that of the A^- ions. Since the salt is completely ionized, the solubility S of the salt can be represented by the individual ion concentration:

$$S = (B^+) = (A^-)$$

By substituting these in Eq. (3.11), K_{sp} can be rewritten as

$$K_{sp} = (B^+)(B^+) = (B^+)^2$$

or

$$K_{sp} = (A^-)(A^-) = (A^-)^2$$

Therefore,

$$(B^+) = (A^-) = \sqrt{K_{sp}} \text{ or } S = \sqrt{K_{sp}} \qquad (3.12)$$

Solubility and solubility products are of significance in systems in which a solid is in equilibrium with its solution. Such systems are common systems in the soil. Dissolution of primary minerals or formation and dissolution of clay minerals are a few examples. In the decomposition of kaolinite, formation of gibbsite obeys solubility laws. The dissolution of liming materials and fertilizers are additional examples. The amount of Ca^{2+} released from the liming material is governed by the solubility product of the component ions. Predictions can be made for the concentration of Ca^{2+} liberated from, and made available to, plant growth by the use of the laws of solubility.

3.10 DISSOCIATION OF WATER

Water molecules have a slight tendency to break up (dissociate) into hydroxyl (OH^-) and hydrogen (H^+) ions. The dissociation is so slight that only one in 10^7 molecules of water is ionized at any one time. However, without this slight ionization, many of the processes and reactions in water would not be possible:

$$H_2O \rightleftharpoons H^+ + OH^-$$

By applying the mass action law, the following relationship is obtained:

$$K_{eq} = \frac{C_{H^+} \times C_{OH^-}}{C_{H_2O}} = 1.8 \times 10^{-16}$$

K_{eq} is called the *dissociation constant* of water. The concentration C of pure water is 55.5 mol/L. By substituting this for C_{H_2O}

$$C_{H^+} \times C_{OH^-} = 1.8 \times 10^{-16} \times 55.5$$
$$K_w = 1.01 \times 10^{-14} \text{ at } 25°C = 298 \text{ K}$$

K_w is called the *ion product* of water and is used for the formulation of pH (see Chapter 9 for discussion).

3.11 DISSOCIATION OF STRONG ELECTROLYTES

Strong electrolytes are dissociated in water completely into their ionic components. NaCl in solid form exists even as Na^+ and Cl^- ions. The HCl in water is usually completely dissociated into H^+ and Cl^- ions. It is common practice to present the dissociation of

HCl as:

$$HCl \rightleftharpoons H^+ + Cl^-$$

However, the reaction is more accurate as follows:

$$HCl + H_2O \rightleftharpoons H_3O^+ + Cl^-$$

H_3O^+ is called a *hydronium ion*.

Now we have seen how a strong electrolyte behaves, let us consider again the reactions using a hypothetical compound HA. The dissociation of HA can be represented as follows:

$$HA \rightleftharpoons H^+ + A^-$$

Applying the mass action, the equilibrium constant, K_{eq}, of the reaction is

$$K_{eq} = K_a = \frac{(H^+)\,(A^-)}{(HA)}$$

Here, K_{eq} is also called K_a, or ionization constant. If $(H^+)\,(A^-) = (HA)$, then $K_a = 1.0$. *Strong electrolytes* can then be defined as electrolytes with $K_a > 1.0$. The stronger the electrolyte, the more complete the dissociation, hence, the larger will be the value of $(H^+)\,(A^-)$. As the K_a value becomes smaller, the rate of dissociation of the electrolytes decreases, and the smaller will be the value of $(H^+)\,(A^-)$. A *weak electrolyte*, therefore, is defined as an electrolyte with $K_a < 1.0$.

When in the foregoing equation $(A^-) = (HA)$ then:

$$K_a = (H^+)$$

Multiplying this equation with $-\log$ gives:

$$-\log K_a = -\log (H^+)$$

or

$$pK_a = pH$$

Hence, the pK_a can be defined as the pH at which the electrolyte dissociates and produces equal concentrations of anions and undissociated molecules. The pK_a value is an important property of chemical compounds, since it indicates their ionization capability. Monoprotic acids have one pK_a value, whereas polyprotic acids (acids capable of dissociating more than one proton per molecule) are characterized by more than one pK_a value. The latter depends on the stage or type of dissociation reaction as illustrated below:

$H_3PO_4 \rightleftharpoons H^+ + H_2PO_4^-$	$pK_{a1} = 2.12$	$K_{a1} = 7.52 \times 10^{-3}$
$H_2PO_4^- \rightleftharpoons H^+ + HPO_4^{2-}$	$pK_{a2} = 7.21$	$K_{a2} = 6.23 \times 10^{-8}$
$HPO_4^{2-} \rightleftharpoons H^+ + PO_4^{3-}$	$pK_{a3} = 12.32$	$K_{a3} = 2.20 \times 10^{-13}$

Consequently, at pH 2.12 ($= pK_{a1}$) the first dissociation is only half complete. At this stage the acid behaves as a strong acid. At pH 7.21 ($= pK_{a2}$), the second proton will be dissociated, and equal concentration of HPO_4^{2-} and $H_2PO_4^-$ ions will be present. The pH must be 12.32 ($= pK_{a3}$) before the third and final dissociation of o-phosphoric acid is 50% completed. At this final stage, H_3PO_4 is considered a weak acid. Therefore, under normal soil conditions, the two predominant ionic species of phosphoric acid will be $H_2PO_4^-$ and HPO_4^{2-}, since soils with a pH $= 12.32$ are exceptions and do not constitute good agricultural soils.

3.12 DISSOCIATION OF WEAK ELECTROLYTES

Weak acids and bases exhibit only a slight dissociation. Acetic acid in water theoretically dissociates into

$$CH_3COOH \rightleftharpoons H^+ + CH_3COO^-$$

By applying the mass action the extent of dissociation can be written as

$$K_{eq} = \frac{(H^+)(CH_3COO^-)}{(CH_3COOH)}$$

In this type of reaction, K_{eq} is called *ionization constant* or K_a. At

25°C the ionization constant is

$$K_a = 1.8 \times 10^{-5}$$

When the concentration of the anion is equal to that of nonionized acid, $K_a = (H^+)$. This is, for example, attained by mixing 0.1 mol sodium acetate with 0.1 mol acetic acid. Under this condition $(CH_3COO^-) = (CH_3COOH) = 0.1$ mol. Therefore,

$$K_a = \frac{(H^+)\,(CH_3COO^-)}{(CH_3COOH)} = 1.8 \times 10^{-5}$$

$$= \frac{(H^+)\,(0.1)}{(0.1)} = 1.8 \times 10^{-5}$$

$$= (H^+) = 1.8 \times 10^{-5}$$

$$pK_a = pH = 5 - \log 1.8$$

$$= 4.74$$

3.13 THE HENDERSON–HASSELBALCH EQUATION

The foregoing discussion is applied for describing ionic properties of amino acids by Henderson–Hasselbalch. If we consider a weak acid HA, then according to the foregoing, the ionization constant is

$$K_a = \frac{(H^+)\,(A^-)}{(HA)}$$

Rearranging this equation gives

$$(H^+) = K_a \frac{(HA)}{(A^-)}$$

By taking the log, the equation is transformed into

$$\log (H^+) = \log K_a + \log \frac{(HA)}{(A^-)}$$

Multiplication with -1 gives

$$-\log (H^+) = -\log K_a - \log \frac{(HA)}{(A^-)}$$

$$pH = pK_a + \log \frac{(A^-)}{(HA)} \tag{3.13}$$

Equation (3.13) is called the *Henderson–Hasselbalch equation* and is used to predict the behavior of ampholytes, such as amino acids and proteins, in solution. All amino acids contain ionizable groups that act as weak acids or bases and give or take protons with change in pH. The ionization of such amphoteric compounds follows the Henderson–Hasselbalch equation, which can be written in the following generalized form:

$$pH = pK_a + \log \frac{\text{Unprotonated Form (Base)}}{\text{Protonated Form (Acid)}} \tag{3.14}$$

When the ratio of the concentration of the unprotonated form to that of the protonated form equals 1, the entire log expression becomes zero. Hence pH $= pK_a$, as shown earlier for weak acids. Consequently, pK_a can be defined (considered synonymous) as the pH when the concentration of unprotonated and protonated species are equal. The pK_a also equals pH when the ionizable group is at its best buffering capacity.

Application of the Henderson–Hasselbalch Concept

Assume again the dissociation of a hypothetical organic acid, HA:

$$HA \rightleftharpoons H^+ + A^-$$
$$1 - a \quad a \quad a$$

where a = fractional amounts in cations and anions dissociated, and $1 - a$ = remaining amounts of undissociated molecules at equilibrium.
Application of the mass action law gives:

$$K = \frac{(H^+)(A^-)}{(1 - a)} \tag{3.15}$$

Rearranging Eq. (3.15) yields:

$$H^+ = K \frac{(1 - a)}{(A^-)}$$

Since $(A^-) = a$, the equation can be written as:

$$H^+ = K \frac{(1 - a)}{(a)}$$

By taking the $-\log$, the equation is transformed into:

$$-\log H^+ = -\log K - \log \frac{(1 - a)}{(a)} \quad \text{or}$$

$$pH = pK + \log \frac{(a)}{(1 - a)} \tag{3.16}$$

Equation (3.16) is valid for the dissociation of monoprotic acids, as indicated by the foregoing dissociation reaction. For polyprotic acids for which the dissociation yields a number of n anions, the equation is formulated as:

$$pH = pK + n\log \frac{(a)}{(1 - a)} \tag{3.17}$$

If $n = 1$, Eq. (3.17) returns to Eq. (3.16). For polyprotic acids, $n = 2$ or more.

Equation (3.17) is sometimes called the *modified* Henderson–Hasselbalch equation. It represents a linear regression curve, in which pK is the intercept, and n is the slope of the curve.

3.14 THE EQUILIBRIUM CONSTANT AND ION PAIRS

Ion pairs are defined as pairs of oppositely charged ions that behave as a thermodynamic entity (Davies, 1962). Strong electrolytes often will not dissociate completely into their component ions. Because of short-range interactions between closely adjacent cations and anions, these ions remain strongly attracted to each other. Therefore, a considerable portion may behave as if they were not ionized at all.

Pairing of ions can be illustrated as follows with Ca^{2+} and CO_3^{2-}:

$$CaCO_3^0 \rightleftharpoons Ca^{2+} + CO_3^{2-}$$

in which $CaCO_3^0$ is the $Ca^{2+}-CO_3^{2-}$ pair. The equilibrium constant for such a reaction is

$$K_{eq} = \frac{(Ca^{2+})(CO_3^{2-})}{(CaCO_3^0)}$$

Although K_{eq} is formulated in a way similar to that of weak electrolytes, the dissociation of ion pairs is affected by forces different from those in weak electrolytes. The attraction in ion pairs is caused by coulombic forces, whereas in weak electrolytes, covalent bonds are the reasons for a weak dissociation.

Soil cations and anions that have been reported to pair extensively are H^+, K^+, Na^+, Ca^{2+}, OH^-, HCO_3^-, CO_3^{2-}, and SO_4^{2-}. Chloride, Cl^-, ions do not form ion pairs with other cations to any measurable amounts (Garrels and Christ, 1965, Davies, 1962).

3.15 THE EXCHANGE CONSTANT AND ION EXCHANGE

Negatively charged organic and inorganic soil colloids have the capacity to adsorb and exchange cations, a topic that will be discussed in detail in Chapter 7 on cation exchange. An example of an exchange reaction involving monovalent ions can be illustrated as follows:

$$\boxed{micelle}\,A + B^+ \rightleftharpoons \boxed{micelle}\,B + A^+$$

Application of the mass action law gives

$$K_{ex} = \frac{(A^+)[B^+]}{[A^+](B^+)}$$

where

( ) = free ion activities
[ ] = adsorbed ion activities
K_{ex} = exchange constant

Rearranging the formula shows the following relation:

$$\frac{(A^+)}{(B^+)} = K_{ex} \frac{[A^+]}{[B^+]}$$

For monovalent cation exchange, the exchange constant K_{ex} is considered a measure of the selectivity of the exchanger or the micelle (clay particle). The *ion selectivity* is defined as the tendency to adsorb one ion more strongly than another. Suppose $K_{ex} = 5$ and the ratio $(A^+)/(B^+)$ of the ion activities in solution equals 1, then one of the ions will be adsorbed more strongly by the micelle than the other. Equal amounts of adsorption of A^+ and B^+ ions take place only if the activity ratio of the ions in solution has the same value as the exchange constant K_{ex}.

3.16 RELATIONSHIP BETWEEN EQUILIBRIUM CONSTANT AND CELL OR ELECTRODE POTENTIAL

If the equilibrium reaction $A + B \rightleftharpoons C + D + ne$ is considered, then in accordance to the concepts of the oxidation potentials the following relationship is valid:

$$E_h = E^o + \frac{RT}{nF} \ln \frac{(C)(D)}{(A)(B)}$$

Since

$$\ln \frac{(C)(D)}{(A)(B)} = \ln K_{eq}$$

therefore,

$$E_h = E^o + \frac{RT}{nF} \ln K_{eq} \tag{3.18}$$

where

E_h = oxidation potential
E^o = standard oxidation potential

R = gas constant
T = absolute temperature
n = valence
F = Faraday constant
K_{eq} = equilibrium constant

Changing from the natural to the common logarithm ($\ln$ = 2.303 log), E_h assumes the following formula:

$$E_h = E^o + \frac{0.059}{n} \log K_{eq}$$

3.17 EQUILIBRIUM CONSTANT AND FREE ENERGY RELATIONSHIP

The derivation of the laws governing equilibrium constants comes from thermodynamics. Chemical thermodynamics is the science of energy relations within chemical systems. In any chemical reaction, energy changes are occurring. A system that is not in equilibrium will spontaneously undergo changes by releasing energy. At equilibrium, the energy changes of the reactants must equal the energy changes of the products, and the following relationship is valid:

$$\Delta G_r = \sum \text{Free Energy Products} -$$
$$\sum \text{Free Energy Reactants} = 0 \quad (3.19)$$

Equation (3.19) expresses the first law in thermodynamics; ΔG_r is the free energy change of reaction. The use of the symbol G is preferred by many authors over the symbol F since G denotes the Gibbs free energy.

For a general reaction A + B $\rightleftharpoons$ C + D, the free energy change of reaction in thermodynamics is written as

$$\Delta G_r = \Delta G_r^0 + RT \ln \frac{(C)(D)}{(A)(B)}$$

or

$$\Delta G_r = \Delta G_r^0 + RT \ln K \quad (3.20)$$

where

K = the activity ratio
ΔG_r^0 = standard free energy change of reaction

If ΔG_r has a negative value, the reaction will go spontaneously to the right. However, if ΔG_r is positive, the reaction will occur in the reverse direction or to the left. If, on the other hand, $\Delta G_r = 0$, then in accordance to the first law of thermodynamics, the reaction is at equilibrium. At equilibrium the activity ratio equals the equilibrium constant K_{eq}. K_{eq} is also known as the thermodynamic equilibrium constant. Consequently, at equilibrium condition the following relation is valid:

$$\Delta G_r = 0 = \Delta G_r^0 + RT \ln K_{eq}$$

therefore

$$\Delta G_r^0 = - RT \ln K_{eq} \tag{3.21}$$

or

$$\Delta G_r^0 = - 1.364 \log K_{eq} \qquad 25°C = 298 \text{ K}$$

3.18 EQUILIBRIUM CONSTANT AND ELECTRON ACTIVITY

As discussed earlier, electrons are active components in a reaction, as with H^+ ions. Consequently, the number of electrons transferred in a reaction can be included in the derivation of the equilibrium constant K. Consider again the generalized half-cell redox reaction:

$$\text{Oxidation} + n \text{ e}^- \rightleftharpoons \text{Reduction} \tag{3.22}$$

This equation indicates that, in analogy to the acid–base concept of Brønsted (see Chapter 9), compounds capable of performing oxidation are then *electron acceptors*, whereas those causing reduction are *electron donors*.

At standard state, the equilibrium constant of Eq. (3.22) is:

$$K^0 = \frac{\text{Reduction}}{(\text{Oxidation}) \, (\text{e}^-)^n} \tag{3.23}$$

By taking the log, the equation is transformed into:

$$\log K^0 = \log \frac{\text{Reduction}}{(\text{Oxidation}) \, (e^-)^n}$$

or $\quad \log K^0 = \log \dfrac{\text{Reduction}}{\text{Oxidation}} + npe$ $\qquad$ (3.24)

The redox potential of Eq. (3.22) is

$$E_h = E^0 - \frac{RT}{nF} 2.3 \log \frac{\text{Reduction}}{\text{Oxidation}}$$ $\qquad$ (3.25)

or $\quad E_h = E^0 - \dfrac{0.059}{n} \log \dfrac{\text{Reduction}}{\text{Oxidation}}$

From Eq. (3.24):

$$\log \frac{\text{Reduction}}{\text{Oxidation}} = \log K^0 - npe$$

Substituting this in Eq. (3.25) gives

$$E_h = E^0 - \frac{0.059}{n} (\log K^0 - npe)$$ $\qquad$ (3.26)

From thermodynamics

$$nFE^0 = RT \ln K^0$$

$$E^0 = \frac{RT}{nF} \ln K^0 \qquad \text{or} \qquad E^0 = \frac{0.059}{n} \log K^0$$

Substituting this in Eq. (3.26) gives

$$E_h = \frac{0.059}{n} \log K^0 - \frac{0.059}{n} (\log K^0 - npe)$$

or

$$E_h = 0.059 \, pe$$ $\qquad$ (3.27)

In the foregoing equation E_h is in volts. As discussed earlier, Eq. (3.27) indicates that the redox potential can be expressed either as

E_h or *pe*, since *pe* can be converted into E_h or vice versa. It is sometimes suggested to express redox conditions in terms of *pe* rather than in terms of E_h, because low *pe* values indicate the presence of reducing conditions or the tendency of the compound to be an electron donor. It then goes without saying that high *pe* values suggest the presence of oxidation conditions or electron acceptors.

3.19 ACTIVITY AND STANDARD STATE

Activity is a measure of the effective concentration of a reactant or product in a chemical reaction. The concentration of a substance does not always accurately describe its reactivity in a chemical reaction. The activity or effective concentration differs from the actual concentration because of interionic attraction and repulsion. The difference between activity and concentration becomes substantially large when the concentration of the reactants is large. At high concentrations, the individual particles of the reactants may exert a mutual attraction to each other, or exhibit interactions with the solvent in which the reaction takes place. On the other hand, in very dilute condition, interactions are less, if not negligible. To correct for the difference between actual and effective concentration, the activity coefficient (γ) is introduced. The activity coefficient expresses the ratio of activity to concentration:

$$\frac{a_A}{c_A} = \gamma \quad \text{or} \quad a_A = \gamma c_A$$

where

γ = activity coefficient
a_A = activity of species A
c_A = concentration of substance A

The activity coefficient is not a fixed quantity, but varies in value depending on the conditions. In very dilute (infinite dilution) conditions, the activity coefficient approaches unity. The value of $\gamma \sim$ 1.0 and, hence, activity equals concentration:

$$a_A = c_A$$

Activity coefficients apply to cations as well as to anions:

Cations: $\gamma_+ = \dfrac{a^+}{m^+}$

Anions: $\gamma_- = \dfrac{a^-}{m^-}$

where

m^+ = concentration of cations
m^- = concentration of anions

The mean ionic activity coefficient is then

$$\gamma_\pm = \left[\left(\frac{a^+}{m^+}\right)\left(\frac{a^-}{m^-}\right)\right]^{1/2} \quad or \quad \gamma_\pm = [(\gamma_+)(\gamma_-)]^{1/2}$$

The mean ionic activity formulated is valid only for monoprotic (1–1) electrolytes, such as HCl and NaCl.

For polyprotic electrolytes, the formula for the mean activity coefficient of the ions changes into:

$$\gamma_\pm = [(\gamma_+)(\gamma_-)^2]^{1/3} \quad \text{for compounds such as } CaCl_2$$

or

$$\gamma_\pm = [(\gamma_+)^2(\gamma_-)]^{1/3} \quad \text{for compounds such as } H_2SO_4$$

Consequently, the mean activity coefficient of the ions in "1–3" and "3–1" electrolytes (e.g., $AlCl_3$) is then

$$\gamma_\pm = [(\gamma_+)(\gamma_-)^3]^{1/4} \quad \text{for compounds such as } AlCl_3$$

or

$$\gamma_\pm = [(\gamma_+)^3(\gamma_-)]^{1/4} \quad \text{for compounds such as } H_3PO_4$$

Activity coefficient in the standard state is indicated by γ^0. A standard state is defined for each substance in terms of a set of reference conditions. Each pure substance in its standard state is assigned an activity of unity. The standard state of solids and liquids is usually chosen as the pure substance under standard conditions of 1 atm pressure and a specified temperature. Since 298.15 K, equivalent to 25.0°C, is a commonly used temperature, it is called the *reference temperature*. The standard state of gases is a perfect gas, obeying $PV = nRT$, at 1 atm pressure and a specified temperature.

3.20 DEBYE–HÜCKEL THEORY AND ACTIVITY COEFFICIENTS

The individual ion activity coefficient can be calculated using the Debye–Hückel equation:

$$-\log \gamma_i = \frac{A \, z_i^2 \, \sqrt{I}}{1 + a_i^0 \, B \, \sqrt{I}} \tag{3.28}$$

where

A, B = constants of the solvents at specified temperature and pressure
z = valence
I = ionic strength
a = effective diameter of the ion

Values of A and B as function of temperature at 1 atm are given as follows:

Temperature		A	B ($\times 10^{-8}$ cm)
0°C	273 K	0.4883	0.3241
5	278	0.4921	0.3249
10	283	0.4960	0.3258
15	288	0.5000	0.3262
20	293	0.5042	0.3273
25	298	0.5085	0.3281
30	303	0.5130	0.3290
35	308	0.5175	0.3297
40	313	0.5221	0.3305

Source: Manov et al. (1943), Garrels and Christ (1965).
Values for a_i^0 can be found in Klotz (1950).

When the ionic strength becomes very small, Eq. (3.28) changes into $-\log \gamma = A\, z_i^2\, \sqrt{I}$.

3.21 IONIC STRENGTH

The concept of ionic strength was introduced by Lewis and Randall (1921) to assess the combined effect of the activities of several electrolytes in solution on a given electrolyte. It is a useful relation in comparing solutions of diverse composition, as in soil water, in river water, and in lake water. The *ionic strength* is defined as

$$I = \tfrac{1}{2}\sum m_i\, z_i^2 \tag{3.29}$$

where

m = moles of ions
z_i = charge of the ions
I = ionic strength

The summation is taken over all ions, positive and negative. For example the ionic strength of a 1 M CaCl$_2$ solution is

$$I = \tfrac{1}{2}\sum[(m_{Ca} \times 2^2) + (m_{Cl} \times 1^2) + (m_{Cl} \times 1^2)]$$
$$= \tfrac{1}{2}\sum[(1 \times 4) + 2(1 \times 1)] = 3$$

That of a $\frac{1}{2}$ M NaCl solution is

$$I = \tfrac{1}{2}\sum[(m_{Na} \times 1^2) + (m_{Cl} \times 1^2)]$$
$$= \tfrac{1}{2}\sum[(\tfrac{1}{2} \times 1) + (\tfrac{1}{2} \times 1)] = \tfrac{1}{2}$$

In some of the physical chemistry books, the formula for the ionic strength is written as

$$I = \tfrac{1}{2} \sum \frac{(m_+ z_+ + m_- z_-)}{m^0} \tag{3.30}$$

where m^0 = moles at standard state = unity or 1. Consequently Eq. (3.29) can be changed into

$$I = \tfrac{1}{2} \sum (m_+ z_+ + m_- z_-)$$

In practice, this definition is similar to the definition as expressed in Eq. (3.29).

Practice

Try to calculate the ionic strength of the following soil solution, for which chemical analysis revealed ion concentrations as listed:

Ion	Parts per million	Moles	$m_i z_i^2$
Na^+	2300	$\dfrac{2300}{23 \times 10^3} = 0.100$	$0.100 \times 1^2 = 0.100$
Ca^{2+}	80	$\dfrac{80}{40 \times 10^3} = 0.002$	$0.002 \times 2^2 = 0.008$
Mg^{2+}	48	$\dfrac{48}{24 \times 10^3} = 0.002$	$0.002 \times 2^2 = 0.008$
SO_4^{2-}	288	$\dfrac{288}{96 \times 10^3} = 0.003$	$0.003 \times 2^2 = 0.012$
Cl^-	1750	$\dfrac{1750}{35 \times 10^3} = 0.050$	$0.050 \times 1^2 = 0.050$
CO_3^{2-}	60	$\dfrac{60}{60 \times 10^3} = 0.001$	$0.001 \times 2^2 = 0.004$
HCO_3^-	2745	$\dfrac{2745}{61 \times 10^3} = 0.045$	$0.045 \times 1^2 = 0.045$

$I = \tfrac{1}{2}(0.100 + 0.008 + 0.008 + 0.012 + 0.050 + 0.004 + 0.045) = 0.114.$

The average ionic strength for water in rocks is about 0.100, whereas streams and lakes have ionic strengths of about 0.010. The ionic strength of ocean waters is approximately 1.0 (Garrels and Christ, 1965) .The higher the ionic strength, the lower will be the ion activity (lower γ), as illustrated by the data in Table 3.2.

Table 3.2 Effect of Ionic Strength on Single-Ion Activity Coefficient

a^a	Ionic strength of solution				
	0.001	0.005	0.01	0.05	0.10
Monovalent ion activity					
3	0.964	0.925	0.899	0.805	0.755
4	0.964	0.927	0.901	0.815	0.770
9	0.967	0.933	0.914	0.860	0.830
Divalent ion activity					
5	0.868	0.744	0.670	0.465	0.380
6	0.870	0.749	0.675	0.485	0.405
8	0.872	0.755	0.690	0.520	0.450
Trivalent ion activity					
4	0.725	0.505	0.395	0.160	0.095
9	0.738	0.540	0.445	0.245	0.180

a^a = effective diameter.
Monovalent, a = 3: K^+, Cl^-, Br^-, I^-, NO_3^-; a = 4: Na^+; a = 9: H^+.
Divalent, a = 5: Sr^{2+}, Ba^{2+}, Ra^{2+}, Cd^{2+}; a = 6; Ca^{2+}, Cu^{2-}, Zn^{2+}, Mn^{2+}, Fe^{2+}; a = 8: Mg^{2+}
Trivalent, a = 4: PO_4^{3-}; a = 9: Al^{3+}, Fe^{3+}, Cr^{3+}.
Source: Klotz (1950).

4

Colloidal Chemistry of Organic Soil Constituents

4.1 THE COLLOIDAL SYSTEM

A *colloid* is a state of matter consisting of very fine particles that approach, but never reach, molecular sizes. The upper size limit of colloids is 0.2 μm, and the lower size limit is approximately 50 Å (5 nm), the size of a molecule. Colloidal systems can be divided into two groups. The colloid is considered *lyophobic* (solute hating) if the dispersed phase does not interact with the dispersion medium. It is called *lyophilic* if it does interact. If the dispersion medium is water, often the terms *hydrophobic* and *hydrophilic* may be used. A hydrophobic colloid can be flocculated, but a hydrophilic usually cannot. Many organic compounds exhibit both hydrophobic and hydrophilic characteristics in the same molecule. Such compounds or molecules are called *amphiphilic*, such as phospholipids and many amino acids. Detergents are examples of synthetic amphiphilics. In some of the amphiphilics, one end of the molecule is hydrophobic, whereas the other end is hydrophilic. These compounds are good surfactants.

The colloids can exist as a gel or a sol. A *gel* is defined as a colloid that exhibits the properties of a solid. Two types of gels are recognized: (1) *True gels*, which are formed by coagulation of lyophilic colloids—fruit jellies and gelatins are good examples of true gels; and (2) *gels*, which are precipitates formed by the coagulation of lyophobic colloids. They are sometimes referred to as *coagula*. *Sols*, on the other hand, are colloids that behave as liquid. A sol exists as a fluid and, to the eye, it is homogeneous and looks like a solution. If the dispersion medium is water, the terms *hydrosols* and *hydrogels* are often used.

Plant and soils contain large amounts of solid material that is in a colloidal state. Such material exhibits chemical and physical properties that depend on the colloidal condition. The inorganic fraction of soils is made up of boulders, rocks, gravel, sand, silt, and clay. Clay comprises all inorganic solids smaller than 0.002 mm (<2 μm) in effective diameter and is considered a soil colloid. Soil organic matter and plant solids also occur in the colloidal state. Humus, protoplasm, and cell walls exhibit many properties of colloidal systems.

Many chemical and biological reactions occur at the solid–liquid interfaces. *Adsorption* takes place at the interface. This refers to the concentration of materials at the colloidal surface. In contrast, *absorption* indicates the uptake and retention of one material within another. Sometimes, it is difficult to distinguish between adsorption and absorption, and in such a case the term *sorption* is proposed. *Desorption* is used to indicate the release or removal of materials that were adsorbed. The substance sorbed is called the *sorbate*, and the material in which sorption occurs is called the *sorbent*.

4.2 THE ORGANIC COMPONENTS

The organic components of soils originate from the biomass that is characteristic for an active soil. Although, strictly speaking, both living organisms and the dead components are included in soil organic matter, only the nonliving fraction will be discussed here. The nonliving organic components are formed by chemical and biological decay of mainly plant materials. They can be divided into (1) materials in which the anatomy of the plant substance is still visible and (2) completely decomposed materials. The first group is of significance in soil physics (e.g., protection of soils by leaf mulch, de-

creasing bulk density, or soil structure). However, from the standpoint of soil chemistry, the nondecomposed organic fraction is
chemically of minor importance because its intact structure exhibits
a relatively small surface area, rendering it inactive as an adsorbent. Also of major importance are the decomposition products, although their nature and accumulation in soils depend on the types
and quantity of plant material subjected to decomposition.

The plant tissue is composed of C, H, O, N, S, P, and a number
of other elements. The inorganic ions make up the ash content that
sometimes accounts for as much as 10% of the dry weight of the
tissue. The organic part of the plant tissue is composed of numerous
organic compounds, but only a few are present in detectable amounts
in soils after decomposition. They are primarily (1) carbohydrates,
(2) amino acids and proteins, (3) lipids, (4) nucleic acids, (5) lignins,
and (6) humus.

4.3 CARBOHYDRATES

Carbohydrates are, by definition, polyhydroxy aldehydes, ketones,
or substances that yield one of these compounds on hydrolysis. Glucose ($C_6H_{12}O_6$) and fructose ($C_6H_{12}O_6$) are examples of an aldose
and a ketose, respectively:

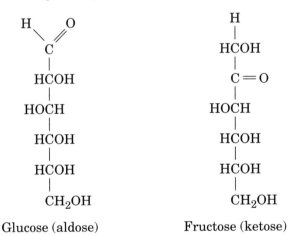

Glucose (aldose)   Fructose (ketose)

The aldose group is characterized by a terminal carbonyl, C=O,
group, whereas the ketose group has its carbonyl group in the carbon

chain. The foregoing structure is called an open-chain structure. Originally all sugars were considered as open-chain–structured compounds. However, to explain mutarotation, a ring (cyclic) structure, as proposed by Tollens in 1883, must be present in sugar molecules. The formation of a ring structure can be illustrated as follows:

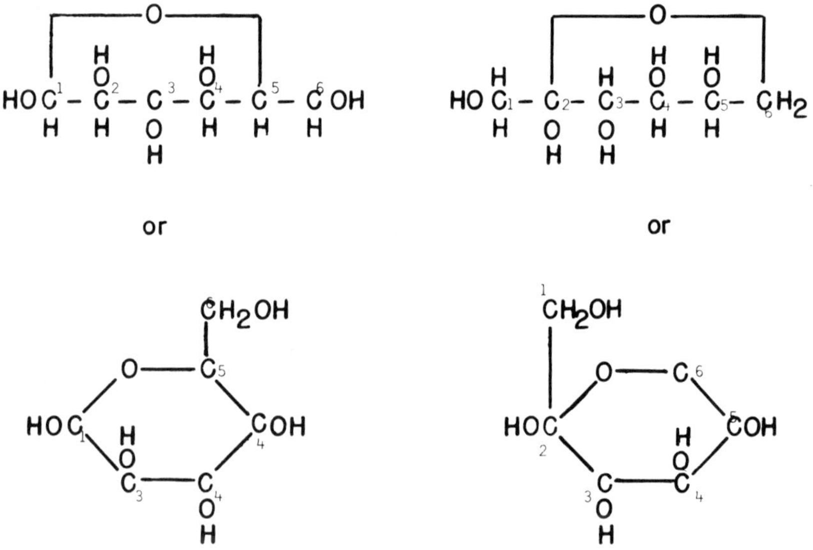

Glucose

Fructose

It is now known that in aqueous sugar solutions the two forms of structures are present, and an equilibrium exists between the forms with cyclic and open-chain structures.

The term *carbohydrate* indicates that these compounds could be represented by hydrates of carbon: $C_x(H_2O)_y$. However, it was found that this definition was not suitable, since several compounds exist with the properties of carbohydrates, but do not have the required ratio of hydrogen to oxygen of $2:1$. The sugar deoxyribose ($C_5H_{10}O_4$), which is a constituent of deoxyribonucleic acid, a component of every plant cell, is an example. Some of the carbohydrates may also contain N and S, and their formulas do not agree with $C_x(H_2O)_y$.

Carbohydrates can be divided into three groups: (1) monosaccharides, (2) oligosaccharides, and (3) polysaccharides. *Monosaccharides* are simple sugars that cannot be hydrolyzed into smaller molecules under reasonably mild conditions. According to the number of carbon atoms, monosaccharides may be trioses ($C_3H_6O_3$), tetroses, and so on up to octoses or nonoses. *Oligosaccharides* are compound sugars that, upon hydrolysis, yield two to six molecules of simple sugars. A disaccharide, for example, hydrolyzes into two monosaccharides and, upon hydrolysis, pentosaccharides yield five monosaccharides. *Polysaccharides* are groups of compounds that yield many different monosaccharides upon hydrolysis. They include cellulose and hemicellulose. Some of the monosaccharides that are bonded together by glucosidic bonds to form polysaccharides are glucose, xylose, and arabinose.

The properties of these carbohydrates change significantly with increasing molecular complexity. The simple sugars are readily soluble in water. The oligosaccharides are crystalline compounds, readily soluble in water, usually having a sweet taste. Polysaccharides are frequently tasteless, insoluble in water, and amorphous. They have high molecular weights. The plant starches and animal glycogens are important examples of polysaccharides. In water they exhibit imbibitions or become dispersed, but they are not strictly soluble. Cellulose, another polysaccharide, is insoluble in water. The molecular weight of cellulose varies from 200,000 to 2 million. Polysaccharides are sometimes divided into homo- and heteropolysaccharides. Homopolysaccharides are composed of a repeating monosaccharide, whereas heteropolysaccharides are made up of two (or more) different monosaccharides. The monosaccharide molecules can be bonded in a straight chain, or they can form branchlike structures (Figure 4.1). Hydrolysis in an acid medium is usually employed to release the monosaccharides. The sugars released are identified either by gas chromatography or by paper chromatography.

Soil polysaccharides may be different from the original plant polysaccharides. They are subject to decomposition by microbial attack, since they are sources of food and energy. Enzymatic attack involves transformation through glycosyl transfer. Two broad types of enzymes have been reported: endo- and exoenzymes. The endoenzymes effect the catalytic cleavage of the glucosidic bonds, whereas the exoenzymes induce the cleavage of terminal residues. The greater the different types of linkage and the greater the branching of the polysaccharide structure, the greater will be the resistance to en-

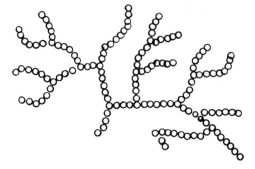

Figure 4.1 Glycogen, a branched-structured polysaccharide from animal tissue. Each circle represents a glucose molecule in the chain.

zymatic degradation of soil polysaccharides. This resistance is perhaps why these compounds can accumulate in soils, although the amount rarely accounts for more than 20% in the soil.

Soil polysaccharides can also be protected against degradation by interaction with other soil constituents, such as clay and metal cations. The intimate association with soil clays has been reported to slow down chemical degradation, whereas adsorption of polysaccharides by, especially expanding, clays [e.g., montmorillonite (Olness and Clapp, 1973, 1975)] in intermicellar spaces renders them inaccessible to enzymatic or to other microbial attack. Evidence has also been presented that complex reaction with metal cations (e.g., Cu, Fe, and Zn) may inhibit enzymatic decomposition of soil polysaccharides (Martin et al., 1966).

Soil polysaccharides influence soil physical conditions, cation exchange reactions, retention of anions, carbon metabolism, biological activity, and complex reactions of metals. They also react with lignin and amino acids and, therefore, contribute toward the formation of humus, humic acids, and related compounds. Mention has been made in the literature that interaction of soil polysaccharides with soil particles encourages soil aggregation, with the consequent formation of granular to crumb structures (Greenland et al., 1961, 1962; Baver 1963). Baver (1963) also indicated that the oxidative destruction of soil polysaccharides resulted in a 30–90% reduction of stability of soil aggregates. This stabilizing effect on soil structure is attributed to an increase in cementation effect. By interaction with soil clays, the polysaccharide is thought to change the properties of the clay surfaces for adsorption of water. The organic compounds compete with water molecules for adsorption sites and reduce wetting and swelling, thereby increasing cementation. The stabilization effect of soil aggregates by fungal mycelia, as frequently postulated by several authors, is considered a temporary effect by Baver (1963), since mycelia and cells undergo further microbial decomposition.

Environmental Importance of Carbohydrates

Carbohydrates are perhaps the most important constituents of plants. They belong to the three major groups of food substances, carbohydrates, proteins, and oil, that plants produce. In fact, carbohydrates are synthesized first by a process called photosynthesis, and the production of the other substances then begins. Accordingly, carbohydrates are the direct link between the radiant energy of the sun and the energy exhibited by living organisms. They are the principal foodstuffs and the most important sources of energy for animals and human beings. Much has been heralded today through the news media about the importance of complex carbohydrates in human nutrition and human health. Ascorbic acid, vitamin C, and inositol are related to carbohydrates (Gortner and Gortner, 1949).

Carbohydrates are renewable resources for raw materials in the pulp, paper, plastic, and rayon industry and, in addition, are used in the production of ethanol, butanol, glycerol, citric acid, acetic acid, and many other chemicals with the assistance of fermentation processes brought about by several microorganisms. In living plants, carbohydrates serve as a source of energy for many biological functions, and play an important role in the synthesis of nucleic acids, lignin, and other structural components in plant tissue. This will be discussed in other sections of the book. Sugars are the preferred source of materials and are subject to anaerobic and aerobic decomposition processes. In the aerobic process, the sugar is broken down completely into CO_2 and H_2O. In the anaerobic microbial process, the sugar is broken down into methane, CH_4, and CO_2. The reaction is usually represented as follows:

$$C_6H_{12}O_6 \rightarrow 3CH_4 + 3CO_2$$

A partial decomposition is also possible, resulting in the production of ethyl alcohol or methyl alcohol, or both. This process can be illustrated with the reaction:

$$C_6H_{12}O_6 \rightarrow 2C_2H_5OH + 2CO_2$$
$$\text{ethanol}$$

This reaction forms the basis for the production of alternative sources of fuel. In practice, sugars from sugar cane, or starches from corn and potato, $(C_6H_{10}O_5)_5$, fermented with yeast, are the least expensive sources. The production of methane and ethanol is always accompanied with a release of CO_2 .The latter may cause some concern about decreasing the quality of the air in the atmosphere. However, in the presence of abundant plant growth,excess CO_2 will be "filtered" from the atmosphere through absorption by green plants and again be used in photosynthesis for the production of carbohydrates.

4.4 AMINO ACIDS AND PROTEIN

Amino acids are the fundamental structural units of protein. The nitrogen in amino acids occurs as an amino (NH_2) group attached to the C chain. The acid part consists of a terminal C linked to an O atom and an OH group, often written as —COOH. The latter, called *carboxyl group*, exhibits acidic properties, since the H in the OH radical is capable of reacting with bases. The general formula of amino acids may be written as

$$
\begin{array}{c}
NH_2 \\
| \\
C - C - COOH \\
| \\
H
\end{array}
$$

Because the amino group is on the carbon adjacent to the carboxyl group (the α-carbon), amino acids with this general formula are called α-amino acids.

The amino acids obtained from hydrolysis of proteins may be classified into (1) aliphatic, (2) aromatic, and (3) heterocyclic amino acids.

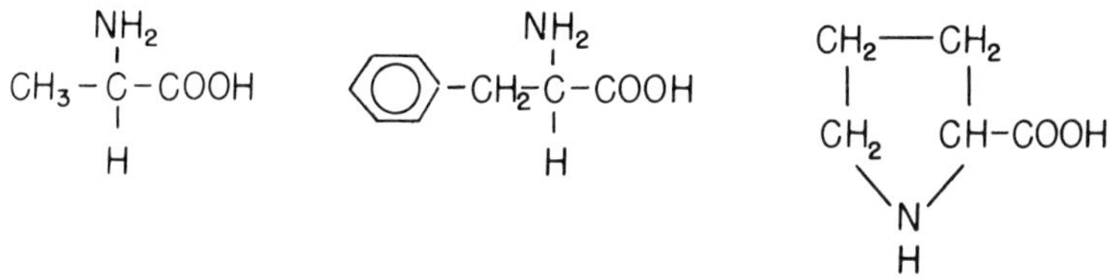

$$CH_3-\underset{\underset{H}{|}}{\overset{\overset{NH_2}{|}}{C}}-COOH$$

$$\langle\bigcirc\rangle-CH_2-\underset{\underset{H}{|}}{\overset{\overset{NH_2}{|}}{C}}-COOH$$

$$\begin{array}{c} CH_2\!-\!CH_2 \\ | \qquad | \\ CH_2 \quad CH\!-\!COOH \\ \diagdown_{\underset{H}{N}}\diagup \end{array}$$

Aliphatic
L-alanine

Aromatic
L-phenylalanine

Heterocyclic
L-proline

Proteins are complex combinations of amino acids. Under refluxing with 6 N HCl for 18–24 hr, the protein may be hydrolyzed to its constituent amino aids. Twenty-one amino acids are usually found as protein constituents. The protein is formed by the linkage of many amino acids through the amino and carboxyl groups:

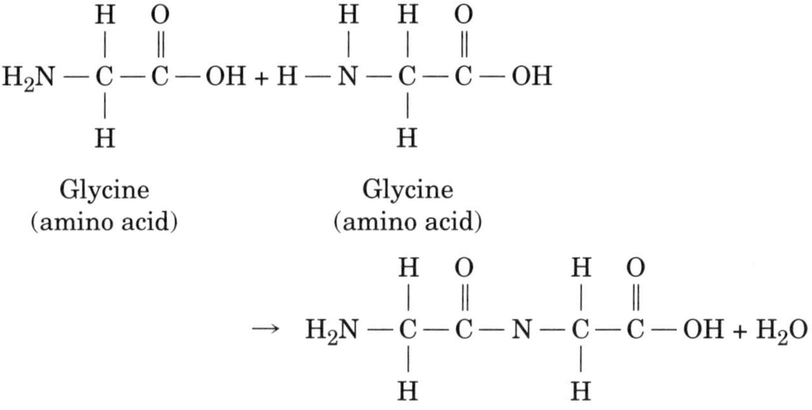

$$H_2N-\underset{\underset{H}{|}}{\overset{\overset{H}{|}}{C}}-\overset{\overset{O}{\|}}{C}-OH + H-\underset{\underset{H}{|}}{\overset{\overset{H}{|}}{N}}-\underset{\underset{H}{|}}{\overset{\overset{H}{|}}{C}}-\overset{\overset{O}{\|}}{C}-OH$$

Glycine
(amino acid)

Glycine
(amino acid)

$$\rightarrow \quad H_2N-\underset{\underset{H}{|}}{\overset{\overset{H}{|}}{C}}-\overset{\overset{O}{\|}}{C}-N-\underset{\underset{H}{|}}{\overset{\overset{H}{|}}{C}}-\overset{\overset{O}{\|}}{C}-OH + H_2O$$

Peptide
(protein)

Since the N content of most proteins is about 16% and since this element is easily analyzed as NH_3 by the Kjeldahl procedure, the protein content can be estimated by determination of the N content and multiplying by 6.25 (100/16).

Zwitterion

The amino acids, with certain exceptions, are generally soluble in water and are insoluble in nonpolar organic solvents such as ether, chloroform, and acetone. Since amino acids contain both a carboxyl and an amino group, these compounds will react with acids and alkalis. Such compounds are said to be amophoteric. If, for example, alanine is dissolved in H_2O, the pH is 7.0. If electrodes are placed in the solution and a potential difference is placed across the electrodes, the amino acid will not migrate in the electric field. However, if alkali is added to the solution, alanine becomes negatively charged and will migrate to the positive anode. Similarly, when acid is added, alanine becomes positively charged and migrates to the negative cathode. This behavior can be explained by considering alanine a zwitterion (Bjerrum, 1923):

$$
\begin{array}{c}
NH_3^+ \\
| \\
H_3C - C - COO^- \\
| \\
H
\end{array}
$$

At pH 7.0, the amino group is still protonated. When alkali is added, the excess proton on the amino group is neutralized ($pK_a = 9.7$):

$$
\begin{array}{c}
COO^- \\
| \\
H - C - NH_3^+ + OH^- \\
| \\
H
\end{array}
\rightarrow
\begin{array}{c}
COO^- \\
| \\
H_2N - C - H + H_2O \\
| \\
CH_3
\end{array}
$$

When acid is added, the dissociated carboxyl group accepts the proton:

$$
\begin{array}{c}
NH_3^+ \\
| \\
H_3C - C - COO^- + H^+ \\
|
\end{array}
\rightarrow
\begin{array}{c}
NH_3^+ \\
| \\
H_3C - C - COOH \\
|
\end{array}
$$

At pH $= 2.3$, the carboxyl group is half protonated.

From the foregoing, it follows that in acid soils, most of the amino acids will then be present as positively charged compounds.

4.5 LIPIDS

Lipids are heterogeneous compounds of fatty acids, waxes, and oils. The term *lipid* does not imply a particular chemical structure, as with amino acids. The name is used to describe substances that are soluble in fat solvents, such as ether, chloroform, or benzene. Lipids are usually classified into three groups:

1. *Simple lipids:* These include neutral lipids, fats, oils and waxes.
2. *Compound lipids:* Phosphatides, glycolipids, sulfolipids, and terpenoid lipids, including carotenoids, belong to this group.
3. *Derived lipids:* These are lipids derived from hydrolysis of simple and compound lipids. They include fatty acids, alcohols, and sterols. The fatty acids can be unsaturated fatty acids (e.g., oleic acids, $C_{18}H_{34}O_2$) and saturated fatty acids (e.g., palmitic acid, $C_{16}H_{32}O_2$). Palm oil, or coconut oil, is rich in palmitic acids. Cholesterol is an example of a sterol, which upon ultraviolet radiation will form vitamin D.

The basic component of lipids is glycerol, $C_3H_8O_3$, or other alcohols. Glycerol is a trihydroxy alcohol with the following structure:

$$
\begin{array}{ccc}
& H & \\
H & O & H \\
OH-C-C&-&C-OH \\
H & H & H \\
\end{array}
$$

Hydrolysis of fats, by saponification with alkalies, yields glycerol and the salts of fatty acids. The metallic salts of the higher fatty acids are known as soap. Neutral lipids are esters of fatty acids and glycerol, and are composed of one molecule glycerol and three molecules of fatty acids. They are liquid at room temperature. Waxes are esters of fatty acids and other alcohols. Phosphatides contain P, and upon hydrolysis will yield glycerol. Lecithin is a phosphatide. The P-containing lipids are also called *phospholipids*. The molecular structure of phosphatides can be represented as follows:

$$
\begin{array}{l}
C-C-O-R_1 \\
| \\
C-C-O-R_2 \\
| \\
C \\
| \quad\nearrow OH \\
O-P=O \\
\quad\searrow OH
\end{array}
$$

Lipids have limited solubility in water and exhibit a hydrophobic character. Many of the lipids in plant and animals are associated with proteins and carbohydrates (e.g., glycolipids). Membrane lipids are amphiphilic because of the presence of hydrophobic and hydrophilic groups in the same molecule.

4.6 NUCLEIC ACIDS

Each plant and animal cell contains a discrete rounded or spherical body, called the nucleus, which contains nucleic acids. *Nucleic acids*, first isolated in 1869 by F. Miescher, are polymers with high molecular weights. Their repeating unit is a mononucleotide, rather than an amino acid. These acids control the synthesis of enzymes and proteins and are also responsible for the genetic transfer in cell division. Two types of nucleic acids are generally recognized: (1) deoxyribonucleic acid (DNA), a constituent of cell nuclei; and (2) ribonucleic acid (RNA), located in the nucleolus and in the cytoplasmic nuclear membrane, called endoplasmic reticulum. Both DNA and RNA consist of long chains of alternating sugar and phosphate residues. In RNA, the sugar is D-ribose. The sugar in DNA, as the name implies, is 2-deoxyribose.

a-D-ribose   a-2-deoxyribose

In most cells, these nucleic acids are conjugated with proteins to form nucleoproteins. Nucleoprotein containing DNA, a major component of chromosomes, determines genetic heredity. On the other hand, nucleoproteins containing RNA, known as *ribosomes*, are important in protein synthesis. Three groups of ribosomal RNA are identified: (1) rRNA (ribosomal RNA), which is the predominant group, amounting frequently to 80% of the total RNA content; (2) soluble RNA (sRNA), sometimes also called transfer RNA (tRNA); and (3) messenger RNA (mRNA), which usually occurs in low concentration. Soluble RNA carries amino acids to their specific sites on the protein template and, therefore, is considered an amino acid carrier or amino acid adaptor, whereas mRNA acts as the messenger of DNA. During the formation of protein, mRNA directs the linkage of amino acids with tRNA.

As the name implies, a nucleic acid is a component of the cell nucleus. However, it is now established that nucleic acids can also originate in other plant parts. It is common for the plant cell to contain a nucleus, several mitochondria, and chloroplasts. Nucleic acids can also be produced by the latter two plant organelles. However, the largest percentage of nucleic acid is produced in the nucleus, and approximately 10–40% DNA and RNA are formed only in the chloroplast, whereas 1% is in the mitochondria. However, these nucleic acids are not only smaller, but their genetic capability is less than those in the nucleus.

4.7 LIGNINS

Lignin is a system of thermoplastic, highly aromatic polymers, derived from coniferyl alcohol or guaiacyl propane monomers. Plant lignin can be divided into three types of basic monomers: (1) lignin from soft wood, (2) lignin from hardwood, and (3) lignin from grasses, bamboo and palm, or grass lignin (Figure 4.2). These basic monomers form large, complex polymeric molecules, and it is common to find a haphazard structure of lignin in many organic chemistry books. A hypothesis is presented here to show that a systematic arrangement of the basic monomers into lignin is possible. Many of the C atoms are connected to the OH radicals (phenolic hydroxyl groups) in which the behavior of the H is much the same as that in the carboxyl groups of organic acids. An example of a systematic linkage of softwood lignin monomers to form polymeric lignin is shown in Figure 4.3. Such a combination can also take place with the other types of monomers, whereas the linkages can continue in many directions.

The ultimate source for formation of lignin is carbohydrates or intermediate products of photosynthesis related to carbohydrates. The process of conversion of the nonaromatic carbohydrates into substances containing phenolic groups characteristic of lignin is called *aromatization*. The end products of the aromatization process are pyrogallol, hydroxyhydroquinone, phloroglucinol, or a combination thereof (Figure 4.4).

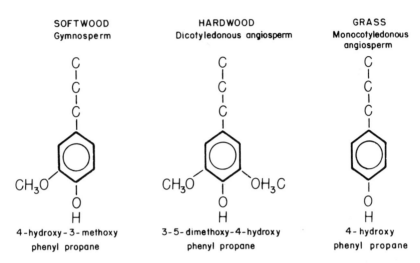

SOFTWOOD
Gymnosperm

HARDWOOD
Dicotyledonous angiosperm

GRASS
Monocotyledonous
angiosperm

4-hydroxy-3-methoxy
phenyl propane

3-5-dimethoxy-4-hydroxy
phenyl propane

4-hydroxy
phenyl propane

Figure 4.2 Chemical structure of building constituents of lignin from softwood, hardwood, and grass.

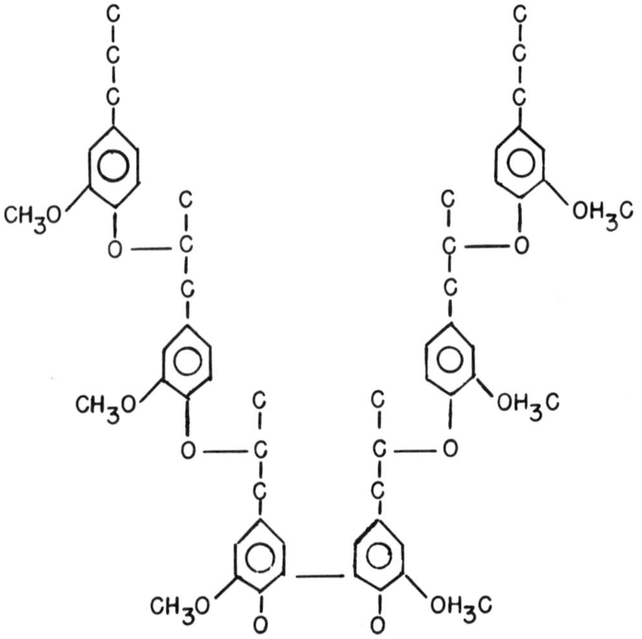

Figure 4.3 Author's hypothesis of a simplified chemical structure of softwood lignin by linkage of coniferyl alcohol monomers. Such a systematic combination is endless.

78

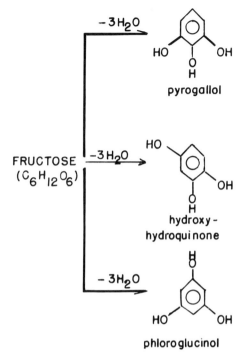

Figure 4.4   Aromatization of carbohydrates through a dehydration process.

In the growth of woody plants, carbohydrates are synthesized first. The formation of lignin then begins, and the spaces existing between the cellulose fibers are gradually filled with lignified carbohydrates. This process is called *lignification* and serves several functions:

1. It cements and anchors the fibers together.
2. It increases the resistance of the fibers against physical and chemical breakdown.
3. It increases rigidity and strength of cell walls.

It is believed that after lignification, the lignified tissue then no longer plays an active role in the life of plants, but serves only as a supporting structure. Nonlignified plant parts contain more moisture, are soft, and break more easily.

The bulk of lignin occurs in the secondary cell walls where it is associated with cellulose and hemicellulose in stems. The quantity of lignin increases with plant age and stem content. It is a very important constituent of woody tissue, and it contains the major portion of the methoxyl content of the wood. A large amount of lignin is also detected in the vascular bundles of plant tissue. The purpose is perhaps to strengthen and make the xylem vessels more water-resistant. By virtue of the presence of larger amounts of vascular bundles, the lignin content of tropical grasses is considerably larger than that of temperate region grasses (Figure 4.5). Consequently, soils under tropical grasses are expected to have higher lignin contents than soils under temperate region grasses. These differences may have some influence on the nature of the humic matter formed.

Lignin is insoluble in water, in most organic solvents, and in

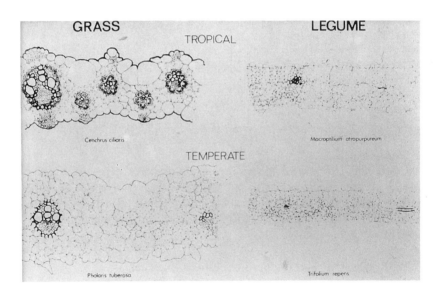

Figure 4.5 Micrograph of leaf thin-sections from tropical and temperate region plants showing different amounts of lignified vascular bundles. Characteristic coloration of lignin was developed by treatment with phenol. *Cenchrus ciliaris* = sandbur grass; *Phalaris tuberosa* = canary, Harding, or Toowoomba grass; *Macroptilium artropurpureum* = siratro twining legume; *Trifolium repens* = white clover. (From D. J. Minson and J. R. Wilson (1980). Courtesy: J. Austr. Inst. Agric. Sci.)

strong sulfuric acid. It has a characteristic ultraviolet (UV) absorption spectrum, and gives characteristics color reactions upon staining with phenols and aromatic amines. It hydrolyzes into simple products, as do the complex carbohydrates and proteins. When oxidized with alkaline benzene, it yields up to 25% vanillin. Lignin is considered an important source for the formation of soil humus, or humic matter. The high resistance of lignin to microbial decomposition is perhaps why it accumulates in soil. It is believed that, depending on the conditions, this could result in the formation of peat, which, in time, can be converted into lignite, coal, and ultimately, oil (fossil fuel) deposits.

4.8 HUMUS AND HUMIC ACIDS

Terminology and Definitions

Soil organic matter is often divided into nonhumified and humified materials. The nonhumified substances are the compounds in plant and other organisms with definite characteristics discussed in the preceding pages (e.g., carbohydrates, amino acids, protein, lipids, nucleic acids, and lignins). These compounds are usually subject to degradation and decomposition reactions. But sometimes they can be adsorbed by inorganic soil components, such as clay, or they may occur in anaerobic conditions. Under such conditions, the foregoing compounds will be relatively protected against decomposition. The humified fraction is known as *humus*, or currently as *humic compounds*, and is considered the end product of decomposition of plant material in soils. The term *humic acid* originated with Berzelius in 1830, who classified the soil humic fraction into (1) humic acid, the fraction soluble in bases; (2) crenic and apocrenic acid, the fraction soluble in water; and (3) humin, the insoluble and inert part. Humic acid was also referred to as ulmic acid, whereas humin was also called ulmin by Mulder in 1840. In 1912, Oden proposed the use of the name fulvic acid replacing the terms crenic and apocrenic acids.

Today *humic compounds* are defined as amorphous, colloidal polydispersed substances with yellow to brown-black color and relatively high molecular weights. Several authors believe that these compounds are heterogeneous in molecular weight, although chemically they may be homogeneous in composition (Felbeck, 1965). Based on solubility in acids and alkalis, the humic compounds can be separated into several humic fractions (Flaig et al., 1975):

Fractions	Alkali	Acid	Alcohol
Fulvic acid	Soluble	Soluble	
Humic acid	Soluble	Insoluble	Insoluble
Hymatomelanic acid	Soluble	Insoluble	Soluble
Humin	Insoluble	Insoluble	Insoluble

According to the German workers, humic acid can be further separated with neutral salt solutions into brown humic acid (soluble in NaCl) and gray humic acid (insoluble in NaCl). In addition to the major fractions, several authors have also reported the isolation of a green humic acid fraction (Kumada and Hurst, 1967).

Types of Humic Matter

Humic compounds, or humic matter, exist not only in soils, but also in streams, rivers, lakes, oceans, and their sediments. They can also occur in lignite or leonardite, coal, and other geologic deposits. These deposits are the sources for the production of commercial humates (Lobartini et al., 1991; Burdick, 1965) that are used as soil amendments. Consequently, three categories of humic matter can be distinguished:

Terrestrial *or* Terrigenous *Humic Matter.*   This is humic matter in soils that comprises mainly lignoprotein complexes. Humic and fulvic acids are major constituents. From the type of the lignin monomer, the group can perhaps be subdivided into:

1. Softwood terrestrial humic matter, which is structurally characterized by coniferyl alcohol.
2. Hardwood terrestrial humic matter, which is composed of sinapyl alcohol monomers.
3. Grass or bamboo terrestrial humic matter, which consists of coumaryl alcohol monomers.

Aquatic *Humic Matter.*   This is humic matter in streams, lakes, and oceans, and their sediments, and is composed mostly of fulvic acids. Humic acid is only a minor constituent. The group can be subdivided into:

1. *Allochthonous* aquatic humic matter, which is humic mater brought from the outside in water. The humic matter is formed in soils and, after formation, is leached or eroded into rivers, lakes, and oceans. Although physical and chemical changes may be induced by the aquatic environment, the nature of the humic matter is still related to soil humic matter, which consists of lignoprotein complexes.
2. *Autochthonous* aquatic humic matter, which is humic matter formed from cellular constituents of indigenous aquatic organisms. In marine sediments, this kind of humic matter consists of carbohydrate–protein complexes (Jackson, 1975; Degens and Mopper, 1975). The source is organic debris from plankton, seaweed, and kelp.

Geologic *Humic Matter.*   This is humic matter in lignite or leonardite and other geologic deposits. It is composed mostly of humic acids. Because of the aging process, most, if not all, of the fulvic acids have apparently polymerized into humic acids.

Agricultural and Industrial Importance of Humic Acids

Humus and humic substances are very important soil constituents. Depending on climatic conditions and cultural practices, the humus content of soils often stabilizes at a fairly definite amount. In the southern region of the United States, with a prevailing subtropical climate, soil humus content seldom exceeds 3.5% (Tan et al., 1975), and the carbon/nitrogen (C/N) ratio usually narrows down in a humification process from a value in excess of 20 for fresh material, to a value of 8–20 for humus. In the semihumid regions of the Mollisols, the humus content stabilizes at values between 6 and 10%.

Together with soil clays, the humic substances are responsible for many chemical activities in soils They enter in complex reactions and may influence the growth of plants indirectly and directly. Indirectly, they are known to improve soil fertility by modifying physical, chemical, and biological conditions in soils. Directly, humic substances have been reported to stimulate plant growth through their effect on metabolism and on other physiological processes. Humic compounds also participate in soil formation and play an important role, especially in the translocation or mobilization of clays, aluminum, and iron, giving rise to the development of spodic and argillic horizons. In industry, humic acids and related compounds find application for use as drilling muds for oil wells and as emulsifiers.

Because of the importance in soil fertility, trials have been conducted to produce humic substances on a large scale for use as soil amendment, soil conditioner, or fertilizer. They have been distributed commercially under the names of "liquid humic acid" and "clod buster."

Extraction and Isolation of Humic Substances

Several methods are available for the extraction and isolation of humic substances from soils. The selection of a suitable extractant is based on two conditions: (1) The reagent should have no effect on changing the physical and chemical nature of the substances extracted, and (2) the reagent should be able to quantitatively remove the humic substances from soils. Over the years many inorganic and organic solvents have been evaluated for their effectiveness in extracting humic compounds (Stevenson, 1965; Schnitzer et al., 1959), usually with mixed results in meeting the two foregoing conditions. Some of the reagents (e.g., dilute bases) can meet the condition set for quantitative removal of humic fractions. However, all of them will have some influence on modifying the physical or chemical properties of the extracted substances (Flaig et al., 1975), and the possibility of creating artifacts still confronts the investigator. Some of the reported inorganic reagents used in extraction are

Acids	Bases and salts
0.1 N HCl	0.1 N NaOH
0.025 N HF	0.5 N NaOH
1% H_3BO_3	0.1 M Na_2CO_3
	0.1 M NaF
	0.1 M $Na_4P_2O_7$, pH 7.0
	0.1 M $Na_4P_2O_7$, pH 9–10
	0.2 M Na_2-EDTA
	0.1 M $Na_2B_4O_7$

Among these reagents, NaOH and $Na_4P_2O_7$ are most widely employed in extraction. Introduced for the first time in 1919 in a generally accepted procedure by Oden, NaOH appears to be the most effective in quantitative removal of humic substances in soils. Since the use of this reagent may induce autooxidation of humic acids, it is usually suggested to conduct extraction with NaOH under an N_2 gas atmosphere. A solution of 0.1 N NaOH is preferred because it is of a milder nature for extraction than 0.5 N NaOH (Pierce and Felbeck, 1975).

Although not as effective as NaOH, $Na_4P_2O_7$ is used frequently for extraction of humic fractions from soils high in sesquioxides content (Kononova, 1961). To increase effectiveness of extraction, a solution with pH 9–10 is recommended. Although occasional reports to the contrary have been noted, the use of 0.1 M $Na_4P_2O_7$ often eliminates the need of decalcifying the soil samples before extraction, as sometimes required with NaOH. In some instances, it has been reported that humic fractions isolated with $Na_4P_2O_7$ exhibited infrared spectra with better resolution than those obtained by NaOH extraction (Tan, 1978c). However, a comparative study on the effectiveness of NaOH and $Na_4P_2O_7$ extraction of humic acids by Orioli and Curvetto (1980) yielded indications that with the pyrophosphate method three high-molecular-weight fractions of humic acids were not extracted.

Extraction with acids as proposed by Schnitzer et al. (1959) technically yields only fulvic acids, since, by definition, only fulvic acids are soluble in acids.

The organic solvents used for extraction of humic substances were oxalic acid, formic acid, phenol, benzene, chloroform, or mixtures of these, acetylacetone, hexamethylenetetramine, dodecylsulfate, and urea (Schnitzer and Khan, 1972).Thus far, none of these has been satisfactory. By using 0.5 M and 0.1 M hydroxymethylamine, Orioli and Curvetto (1980) obtained humic acids with carboxyl contents different from those extracted with NaOH. However, no differences were noted in the electrophoretograms of both the humic acids.

The most common procedure for extraction and fractionation of humic acids with NaOH is shown in Figure 4.6.

Chemical Characterization and Composition

Elemental Composition.   An example of analytical data of humic and fulvic acids is shown in Table 4.1. Humic acid is usually rich

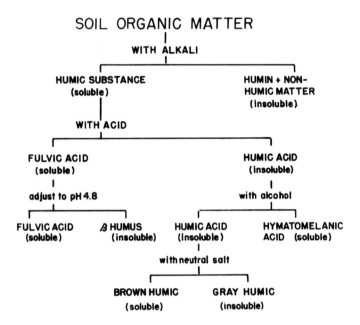

Figure 4.6 Flowsheet for the separation of humic compounds into the different humic fractions.

in carbon, which ranges from 41 to 57%. The lower ranges are exhibited by fulvic acids and humic acids in tropical soils. Fulvic acid distinguishes itself from humic acid also by a higher oxygen, and by lower hydrogen and nitrogen contents. The O content was 44–54% of fulvic acids versus 33–46% in humic acids. The nitrogen content in fulvic acid shows a range of 0.7–2.6% in contrast with humic acids, which contain 2–5% N. For humic acids, Flaig (1975) reported that the brown humic acid fraction of Chernozemic soils (Mollisols) was usually higher in nitrogen content than was the gray humic fraction. By mild degradation of humic compounds with peracetic acid oxidation, Schnitzer and Hindle (1980) succeeded in differentiating the N-containing components in the humic molecule. Different types of N components were detected (e.g., NH_4^+-N, NO_2^- + NO_3^--N, amino acid-N, amino sugar-N, and by difference from total N, "unknown" N. These authors reported that 16.6–59.1% of the unknown N could be converted into NH_3 and N gases.

The amount of N detected in humic acids indicates the necessary

Table 4.1 Chemical Composition of Fulvic (FA) and Humic Acids (HA) Extracted from Tropical and Temperate Region Soils

	C (%)	H (%)	O (%)	N (%)	S (%)	Total acidity (mEq/g)	Carboxyl (mEq/g)	Phenolic hydroxyl (mEq/g)	Alcoholic hydroxyl (mEq/g)	Carbonyl (mEq/g)
HA-Alfisols (temp.)	56.8	5.3	33.3	4.6	nd[a]	6.8	3.9	2.8	nd	nd
HA-Alfisols (trop.)	52.3	5.2	37.2	3.6	nd	nd	nd	nd	nd	nd
HA-Inceptisol (temp.)	51.4	5.8	38.7	4.1	nd	6.0	2.4	3.6	nd	nd
HA-Oxisols (trop.)	44.3	7.7	38.0	2.1	nd	nd	nd	nd	nd	nd
HA-Spodosols (temp.)	56.7	5.2	35.4	2.4	0.4	5.7	1.5	4.2	2.7	0.9
HA-Spodosols (temp.)	49.0	4.6	45.7	0.7	nd	12.0	9.2	2.8	nd	nd
HA-Ultisols (temp.)	48.7	4.8	42.7	3.8	nd	8.7	2.7	6.0	nd	nd
HA-Ultisols (trop.)	44.8	6.3	36.7	2.8	nd	nd	nd	nd	nd	nd
FA-Inceptisol (temp.)	47.9	5.2	44.3	2.6	nd	nd	nd	nd	nd	nd
FA-Spodosols (temp.)	50.9	3.3	44.7	0.7	0.3	12.4	9.1	3.3	3.6	3.1
FA-Spodosols (temp.)	50.2	4.6	43.4	1.8	nd	12.1	7.9	4.2	nd	nd
FA-Ultisols (temp.)	40.6	4.1	53.9	1.4	nd	10.2	8.8	1.4	nd	nd

[a] nd, not determined.

Source: Schnitzer and Khan (1972), Schnitzer (1975), Cranwell and Haworth (1975), Martin et al. (1977), and Tan and Van Schuylenborgh (1959).

participation of N-containing compounds (e.g., amino acids, amino sugar) in the formation of humus. Electron spin resonance (ESR) analysis reveals that humic acids have an aromatic core, containing physically or chemically bonded proteins, polysaccharides, simple phenols, and metals (Cranwell and Haworth, 1975), and according to Ghosh and Schnitzer (1980a), semiquiones also. The linkage to the core renders considerable stability to amino acids, protein, and polysaccharide toward biochemical attack.

From the foregoing data, Kononova (1961) reported that the humus nitrogen content in Chernozems (Mollisols) amounted to 7–10 t/ha in the top 20 cm. This value decreased toward the Spodosols and in Aridisols, where 2–3 t/ha of humus N was reported in the top 20 cm of Sierozems. These figures suggest that humic acid may contribute significantly toward the N supply of soils for plant growth.

Total Acidity. The total acidity or exchange capacity of soil humic compounds is attributed to the presence of dissociable protons or hydrogen ions in aromatic and aliphatic carboxyl and phenolic hydroxyl groups. Humic acids are characterized by a lower total acidity and lower carboxyl content than fulvic acids (see Table 4.1). The total acidity of humic acids amounts generally to 5–6 mEq/g,* with the exception being in humic acids of Spodosols, and Ultisols. Spodosols are very acidic soils containing humic acids that occasionally exhibit a relatively high total acidity and high carboxyl content. In contrast with humic acids, fulvic acids have a total acidity of 10–12 mEq/g, which is approximately two times that of humic acids. The carboxyl content of fulvic acid is two to three times higher than that of humic acids, but the phenolic hydroxyl group concentration does not seem to differ significantly from that of humic acids. Therefore, it can be concluded that the higher acidity of fulvic acid is perhaps attributed to the higher carboxyl content. With the exception noted for one of the Spodosols, the data in Table 4.1 also indicate that the variation observed in total acidity of humic compounds is not affected by the variation in soils, but rather by type of organic acid.

The value for total acidity is, however, also dependent on the method of analysis. It is usually measured by the $Ba(OH)_2$ adsorp-

* *Note*: The unit used by the International System (SI) to express mEq/100 g is cmol (p^+) kg^{-1}.

Table 4.2 Total Acidity Measured by Different Methods

	Total acidity (mEq/g) by			
	$Ba(OH)_2$	KOH	NaOH	Ba acetate
Peat	3.8	2.96	2.54	1.83
Pine Forest soil	6.9	—	—	—
Dunkirk soil (Ontario)	4.3	—	—	—

Source: Felbeck (1953).

tion method, which yields higher values than other methods. As can be noticed from Table 4.2, total acidity measurements using KOH, NaOH, or Ba acetate procedures generally yield lower values (Felbeck, 1965).

Another factor affecting total acidity or exchange capacity of humic substances is the pH. With titration procedures, Tan (1978a) obtained in his chelation studies exchange capacities of humic acids ranging from 2.26 to 4.48 mEq/g. As can be seen from the data in Table 4.3, the total acidity of humic acid was 2.26–2.83 mEq/g at pH 7.0, but increased to 4.43–4.48 mEq/g at pH 11.5. These values were in agreement with those reported by Tiurin and Kononova (1962) and Orlov and Erosiceva (1967). The recovery values for Al^{3+}, Fe^{3+}, and Cu^{2+} of 4.4–4.9 mEq/g HA (pH 11.5) compared favorably with acidity values of humic acids found by Butler [J. H. A. Butler,

Table 4.3 Effect of pH on Amounts of Metal Ions (mEq/100 mg) Chelated by Humic Acids

		pH	
Titration product	Metal ion recovery (mEq/100 mg)	Before titration	After titration
$HA–Al^{3+}$	0.445	11.5	7.05
$HA–Al^{3+}$	0.226	7.0	5.12
$HA–Fe^{3+}$	0.443	11.5	6.42
$HA–Fe^{3+}$	0.255	7.0	4.38
$HA–Cu^{2+}$	0.448	11.5	6.82
$HA–Cu^{2+}$	0.283	7.0	6.03

Source: Tan (1978a).

unpublished Ph.D. thesis, University of Illinois, 1966; see also Stevenson (1976a)].

Carboxyl Groups. Humic acid is generally characterized by a lower carboxyl group content than fulvic acid. The data in Table 4.1 show a carboxyl concentration in humic acids ranging from 1.5 to 2.7 mEq/g HA, in contrast with fulvic acid, with a range of 7.9–9.1 mEq/g FA. The exception is the carboxyl content of humic acid in one of the Spodosols where a value of 9.2 mEq/g HA was detected. As reported earlier, Spodosols are very acidic soils containing acid humus.

Several methods are available for determination of the amounts of carboxyl groups in humic and fulvic acids (e.g., methods using ion exchange, decarboxylation, iodometry, esterification, and the Ca acetate procedure). Many scientists are using the esterification and Ca acetate methods. Depending on the procedures used, different values are obtained. With esterification, employing diazomethane and dimethylsulfate (Tan, 1975; Schnitzer, 1974; Stevenson and Goh, 1972) to esterify the COOH groups, Felbeck (1965) noted that 54% of the exchange capacity of humic acids was attributed to COOH groups.

The Ca acetate method makes use of the reaction in which acetic acid is formed and released according to the reaction:

$$2R-COOH + Ca(CH_3COO)_2 \rightarrow (RCOO)_2Ca + 2CH_3COOH$$

The COOH content is then determined by titration of the acetic acid with a standard base. With this procedure, Schnitzer and Khan (1972) reported a carboxyl content of 1.5 mEq/g of humic acid. As indicated earlier, fulvic acid distinguishes itself from humic acid by containing a higher carboxyl content amounting to 9.1 mEq/g of fulvic acid.

Hydroxyl Groups. Humic substances contain a variety of hydroxyl groups, but for characterization of humic acids generally three types of OH groups are distinguished:

1. Total hydroxyls, measured by acetylation, are the OH groups associated with all functional groups, such as phenols, alcohols, enols, and hydroquinones. However, in many cases the term total hydroxyls refers only to the sum of phenolic and alcoholic-OH groups.

2. Phenolic-OH groups are OH attached to benzene rings. The amount can be calculated by difference as follows:

mEq total acidity − mEq COOH = mEq phenolic-OH

3. Alcoholic-OH groups are OH associated with alcoholic groups. The amount can also be determined by difference as follows:

mEq alcoholic-OH = mEq total OH − mEq phenolic-OH

The amount of phenolic and alcoholic hydroxyl groups does not differ significantly between humic and fulvic acids (see Table 4.1), whereas the reactivity of alcoholic OH groups is usually considered lower than that of phenolic groups.

In addition to the total acidity—carboxylic and hydroxyl groups, as discussed before—humic compounds also contain carbonyl ($C=O$) groups, either as ketonic $C=O$ or as quinoid $C=O$ groups. Each of these groups amounts to approximately only 1 mEq/g of humic or fulvic acid (Schnitzer, 1975). In conclusion it can perhaps be stated that fulvic acid in general, is more acidic and possesses more carboxyl groups than humic acids. Sposito and Holtzclaw (1977) reported that fulvic acid extracted from sewage sludge was a heterogeneous polynuclear polyacid. It contains functional groups that can behave very strongly or weakly acidic. The strongly acidic groups ionize at pH < 2.0 and perhaps are attributable to sulfonic acid groups. The more weakly acidic groups ionize, according to the foregoing authors, at pH > 10.0 and are caused by carboxyls, phenolic-OH, and SH- and N-containing groups. This polyacid nature is considered the reason fulvic acids form weak ion pairs with protons.

Chemical Behavior and Reactions

The chemical behavior of humic matter, in general, is controlled by the two functional groups, the carboxyl- and phenolic-OH groups. The carboxyl group starts to dissociate its proton at pH 3.0 (Posner, 1964), and the humic molecule becomes electronegatively charged (Figure 4.7). At pH < 3.0, the charge is very small, or even zero. At pH 9.0, the phenolic-OH group also starts to dissociate, and the molecule attains a high negative charge. Since the development of negative charges is pH-dependent, this charge is called *variable charge* or *pH-dependent charge*. Several reactions or interactions can

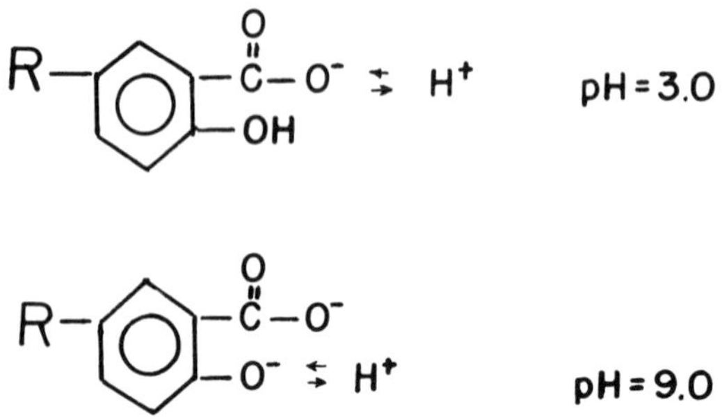

Figure 4.7 Development of variable negative charges in a humic molecule by dissociation of protons from carboxyl groups at pH 3.0 and from phenolic-OH groups at pH 9.0.

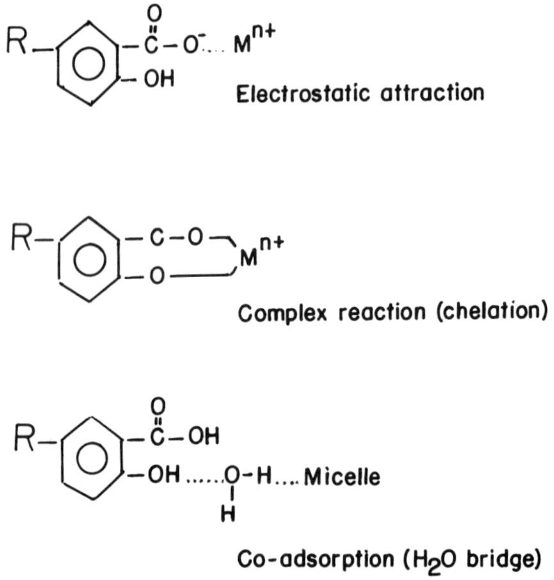

Figure 4.8 Examples of interactions between humic acid and a metal cation, and humic acid and a clay micelle. M^{n+} = cation with charge n^+; R = remainder of humic acid molecule.

take place because of the presence of these charges. At low pH values, the humic molecule is capable of attracting cations, which leads to cation exchange reactions. The cation exchange capacity (CEC) of humic substances can be estimated from their total acidity values. From the data in Table 4.1, it can be noticed that the CEC of humic acids ranges from 500 to 1200 mEq/100 g. Fulvic acids exhibited even higher CEC values. At high pH values, when the phenolic-OH groups are also dissociated, complex reactions and chelation between metals and the humic molecule are of importance (Figure 4.8). Both adsorption and complex reactions can also take place by a water- or metal-bridging process. This happens in interaction reactions between humic matter and clay. In the latter, a water molecule, or the metal, acts as a bridge between the organic ligand (humic molecule) and the clay micelle.

Spectral Characteristics of Humic Compounds

Ultraviolet and Visible Light Spectrophotometry. The color of humic substances is a physical property that has attracted the attention of many scientists who attempted to use it for characterization of humic acid fractions in soils (Flaig et al., 1975; Tan and Giddens, 1972; Schnitzer, 1971; Kononova, 1966; Tan and Van Schuylenborgh, 1961). In Germany, especially, color properties of humic substances have attracted a number of investigators. They reported that the intensity of light absorption was characteristic for the type or molecular weight of humic substances. In UV–visible light spectrophotometry, humic solutions are scanned and the absorbance recorded at various wavelengths between 300 and 800 nm. By plotting the logarithm of the absorbance against the wavelengths, a straight-line regression is obtained. The slope of such a line has been used for differentiation of humic substances. Fulvic acids are considered to yield spectra with a steep slope, in contrast with humic acids (Figure 4.9). The slope of the spectral curve can be expressed as a ratio or quotient of the absorbance at two arbitrarily selected wavelengths (e.g., absorbance at 400 and 600 nm) called the *color ratio*:

$$\text{Color ratio: } E_4/E_6 \text{ or } Q_{4/6} = \frac{\text{extinction (absorbance) at 400 nm}}{\text{extinction (absorbance) at 600 nm}}$$

This color ratio is then used as an index for the rate of light ab-

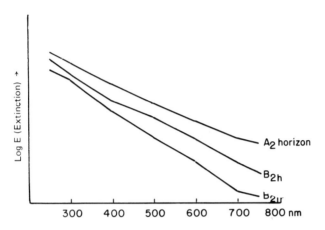

Figure 4.9 Light absorption of humic acids of A and B horizons of a Spodosol in tropical regions. (From Tan and Van Schuylenborgh 1961.)

Table 4.4 Color Ratio E_4/E_6 of Humic Substances Isolated from Soils in Temperate Region

Soil taxonomy orders	Great group	E_4/E_6
	Humic acids, MW $>$ 30,000[a]	
Ultisols	Hapludults (Cecil soil)	4.32
Ultisols	Paleudults (Greenville soil)	4.45
	Humic acids, MW $=$ 15,000[a]	
Ultisols	Hapludults (Cecil soil)	5.49
Ultisols	Paleudults (Greenville soils)	5.47
	Humic acids[b]	
Spodosols	Podzol	5.0
Alfisols	Dark Gray Forest soil	3.5
Mollisols	Chernozem	3.3
Mollisols	Chestnut soil	3.9
Aridisols	Sierozem	4.3
	Fulvic acids	
Ultisols[a]	Hapludults (Cecil soil)	8.0
Unknown (Canada)[b]		6.0–8.0

[a] From Tan and Giddens (1973).
[b] Adapted from Schnitzer and Khan (1972) and Kononova (1966).

sorption in the visible range. A high color radio, 7–8 or higher, corresponds with curves with steep slopes and is usually observed for fulvic acids or humic fractions with relatively low molecular weights. On the other hand a low color ratio, 3–5, corresponds to curves that are less steep. These curves are noted for humic acids and other related compounds with high molecular weights. The data in Table 4.4 show some E_4/E_6 ratios of humic substances extracted from temperate region soils. It can be noticed that humic acids with high molecular weights (MW > 30,000) have lower E_4/E_6 values (4.32–4.45) than humic acids with lower molecular weights (MW = 15,000). The lower-molecular-weight fractions exhibit E_4/E_6 values of 5.47–5.49.

Since UV–visible light spectra of humic compounds are generally featureless lines, Salfeld (1975) proposed the use of differences between two adjacent absorbances. By plotting the logarithms of these differences (ΔE) against the wavelength, a curve was obtained with several peaks (Figure 4.10), called the *derivative spectrum*.

Although the absorption spectra of humic acids in the visible range have the form of straight lines, the inclination or angle of the curves, as expressed by $\Delta \log K = \log K_{400} - \log K_{600}$, has been used in Japan for characterization and distinction of several types of humic acids (Kumada, 1965, Kumada and Miyara, 1973; Yoshida et al., 1978). Four types of humic acid have been recognized:

	$\Delta \log K$	Occurrence
A-type humic acid	<0.6	A horizons of volcanic ash soils
B-type humic acid	0.6–0.8	Brown forest soils, red-yellow soils, and paddy soils
R_p-type humic acid	0.8–1.1	Peat, decomposed grases, stable manure
P-type humic acid	Absorption bands near 615, 570, and 540 nm occur mainly in Podzols (Spodosols). This type of humic acid can be distinguished again in P_g (green) and P_b (reddish-brown) humic acid fractions.	

Finally, it should also be mentioned that humic compounds may exhibit fluorescence spectra. By using fluorescence excitation spec-

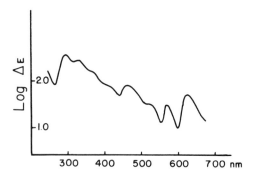

Figure 4.10 Derivative spectrum of humic acid from a Spodosol. (From Salfeld, 1975.)

troscopy, Ghosh and Schnitzer (1980b) indicated that both fulvic and humic acids showed curves with distinct bands at 465 nm. However, fulvic acid appeared to distinguish itself from humic acid by displaying an additional band at 360 nm.

Infrared Spectroscopy. Infrared spectroscopy has been used extensively in the past to characterize humic substances, although some doubt exists on the usefulness of infrared spectra. The latter is perhaps caused in part by the complexity of the infrared spectra of humic preparations and by questions arising about the purity of samples (Tan, 1976b). In spite of these questions, infrared spectroscopy has been very useful and can identify three different types of humic fractions (e.g., fulvic, humic, and hymatomelanic acids) (Figure 4.11; Schnitzer, 1971; Stevenson and Goh, 1972; Stevenson and Goh, 1971; Goh and Stevenson, 1971).

As can be noticed from the spectra, fulvic acid has a strong absorption band at 3400 cm^{-1}, a weak band between 2980 and 2920 cm^{-1}, and a medium strong band at 1720 cm^{-1}, followed by a shoulder at 1650 cm^{-1}, attributed to vibrations of OH, aliphatic C—H, carbonyl (C=O), and carboxyl groups in COO$^-$ form, respectively. The strong band at 1000 cm^{-1} is not necessarily caused by contamination with silica gel (Tan, 1976a,b), but many other functional groups will also absorb in this region, such as ethyl, vinyl —CH=CH$_2$, aromatic aldehyde, amine, and SH groups. The infrared features of fulvic acid as discussed earlier have close similarities with those of polysaccharides (Mortenson, 1961).

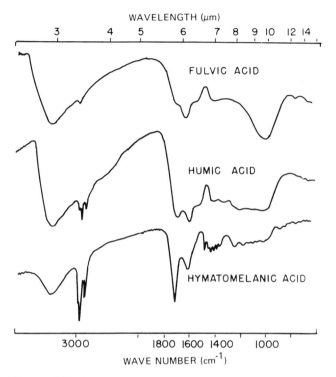

Figure 4.11   Diagnostic infrared characteristics of three major humic compounds, fulvic acid, humic acid, and hymatomelanic acid.

In contrast to fulvic acid, humic acid exhibits a strong absorption for C—H vibrations at 2980–2920 cm^{-1}, and even stronger absorption for both carbonyl and carboxyl vibrations in COO^{-} form at 1720 and 1650 cm^{1-}, respectively. Humic acid spectra have, in addition, no absorption bands at 1000 cm^{-1}.

Hymatomelanic acid, in turn, differs from both fulvic and humic acids by showing infrared spectra with very strong absorption bands for C—H and carbonyl vibrations.

In addition to the functional groups just discussed for humic compounds, a number of other characteristic infrared group frequencies have been detected in organic compounds. This knowledge contributes toward solving the structural chemistry of humic acids, as will be discussed in a later section. The main absorption bands of humic compounds, with their characteristic wave numbers and wavelengths, are listed in Table 4.5.

Magnetic Resonance Spectroscopy. Two types of spectroscopy can be distinguished in this category: electron paramagnetic resonance, which analyzes electron spin resonance of large free radicals in large organic polymers of soil organic compounds; and nuclear magnetic resonance, used for determination of proton resonance in relatively smaller soil organic compounds. These methods have not found wide application because of the expensive instrumentation and the difficulty in using large complex humic compounds.

Electron paramagnetic resonance (EPR), also called electron spin resonance (ESR), analyzes unpaired electron spins in paramagnetic organic materials. The ESR spectra of humic and fulvic acids consist of single lines with hyperfine splitting and Gauss (G) values ranging from 2.0031 to 2.0045, with line widths from 2.0 to 3.6 G (Ghosh and Schnitzer, 1980a; Riffaldi and Schnitzer, 1972). Eltantawy (1980) reported that an enhanced ESR signal (G = 2.0024) indicated the presence of larger free radical contents, but similar G values suggested that the free radicals in humic and fulvic acids were of similar structure.

Analysis of the EPR or ESR spectra by Steelink and Tollin (1967) and Steelink (1964) showed that humic acids exhibited paramagnetism owing to the presence of semiquinones and hydroquinones. An EPR spectrum of a Spodosol (Podzol) and of humic acid extracted from the Spodosol is shown as an example in Figure 4.12. The peak in the Spodosol curves was identified as the organic radical in the humic acid molecule. The spectrum of humic acid supported these observations.

Schnitzer and Khan (1972) reported that if semiquinone is present, this compound can be made to enter in a reaction yielding hydroquinone and quinone:

Semiquinone radical → Hydroquinone + Quinone

Quinhydrone

Table 4.5 Infrared Absorption Bands of Functional Groups in Humic Compounds and Related Organic Substances

Wavenumber, cm^{-1}	Wavelength		Proposed assignments
	μ	nm	
3400	2.94	2940	O—H and N—H stretch
3300	3.03	3030	O—H stretch and N—H stretch
3380	2.95	2950	Hydrogen bonded OH
3100	3.25	3250	NH$_3$ stretch
2985	3.35	3350	CH$_3$ and CH$_2$ stretch
2940	3.40	3400	C—H stretch
2920	3.42	3420	C—H stretch of CH$_3$
2900	3.44	3440	Aliphatic C—H stretch
2820	3.55	3550	CH$_3$ and CH$_2$ stretch
2740	3.65	3650	Hydrogen bonded OH
2610	3.83	3830	Hydrogen bonded OH
2260	4.42	4420	Hydrogen bonded OH
1840	5.43	5430	C=O stretch of cyclic anhydrides, and mixed anhydrides
1815	5.51	5510	C=O stretch of cyclic anhydrides, and mixed anhydrides
1785	5.60	5600	C=O stretch of phenols and cyclic anhydrides
1750	5.71	5710	COOH and C=O
1725	5.79	5790	C=O of carboxyl and C=O of ketonic carbonyl
1720	5.81	5810	C=O stretch of carbonyl groups
1695	5.90	5900	COOH vibrations
1680	5.95	5950	COO$^-$ antisymmetric stretch
1665	6.00	6000	Olefenic C=C
1650	6.0-6.2		C=O stretch (amide I)

Table 4.5 *Continued*

Wavenumber, cm^{-1}	Wavelength		Proposed assignments
	μ	nm	
1630	6.10	6100	Aromatic C=C, hydrogen bonded C=O, double bond conjugated with carbonyl and COO$^-$ vibrations
1613	6.19	6190	C=C and COO$^-$
1610	6.20	6200	Conjugated C=C in ring with C=C, or C=O of open chains
1600	6.27	6270	Aromatic C=C
1590	6.29	6290	Multinuclear aromatic C=C and/or aromatic C=C
1575	6.35	6350	Salts of COOH
1570-1515	6.5-6.6		N-H deformation and C=N stretch (amide II)
1550	6.45	6450	COO$^-$ antisymmetric stretch
1510	6.62	6620	Aromatic C=C
1470	6.80	6800	Aromatic C=C
1460	6.85	6850	CC-H$_3$
1440	6.95	6950	C-H stretch of methyl groups
1435	6.97	6970	C-H bending
1400	7.14	7140	COO$^-$ antisymmetric stretch
1390	7.20	7200	Salts of COOH
1300	7.70	7700	C=N stretch and N-H deformation (amide III)
1280	7.80	7800	C-O stretch
1267	7.89	7890	Aromatic C-O
1230	8.10	8100	C-O ester linkage and phenolic C-OH

Wavenumber, cm^{-1}	Wavelength		Proposed assignments
	μ	nm	
1170-950	8.50-10.5	8500-10500	C−C, C−OH, C−O−C typical of glucosidic linkages, polymeric substances and Si−O impurities in humic compounds
1035	9.67	9670	O−CH$_3$ vibrations
840	11.9	11900	Aromatic C−H vibrations

Source: Adapted from Mortenson et al, (1965), Felbeck (1965), and Schnitzer and Khan (1972).

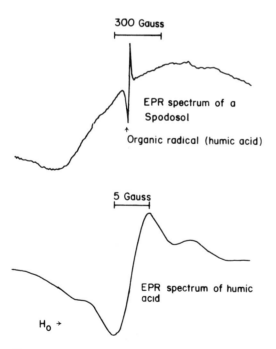

Figure 4.12 EPR spectra of a Spodosol (top) and humic acid extracted from the Spodosol. (From Steelink and Tollin, 1967.)

The quinone released can be easily detected. However, chemical and spectroscopic analyses by the latter authors failed to show measurable quantities of quinone in solutions of degradation products of humic material. This led Schnitzer and Khan (1972) to believe that evidence presented by Steelink and Tollin (1967) for the presence of quinone groups in humic molecules was not convincing. However, Ghosh and Schnitzer (1980a) confirmed by ESR analysis the presence of semiquinone radicals in humic substances. The free radical content in humic and fulvic acids was estimated to range from 1.4 $\times$ 10^{-17} to 37.4 $\times$ 10^{-17} spins/G (Ghosh and Schnitzer, 1980a; Senesi and Schnitzer, 1977).

Nuclear magnetic resonance (NMR), the second method in the category of magnetic resonance spectroscopy, analyzes the hydrogen atoms, or proton resonance, of the humic molecules. Different types of protons in the unknown structure can be detected. However, the usefulness of NMR analysis in humic acid research is questioned by many authors. Nuclear magnetic resonance analysis requires the sample to be dissolved in a suitable solvent; the solvents frequently used were CCl$_4$ and CDCl$_3$. Humic acid, however, is usually not soluble in these reagents and must be first methylated or broken down into smaller molecules by degeneration procedures. Solid samples cannot be used, since they interfere with magnetic interaction. Another possible solvent for NMR analysis is D$_2$O, which finds application with analysis of fulvic acids.

In addition to the foregoing, difficulties also arise because NMR analysis makes use of radio waves, which are low energetic forms of electromagnetic radiation. The level of energy was considered very small, but great enough to affect the nuclear spin of atoms in the nuclei of the poorly defined complex polymers of humic acid molecules.

With use of methylated fulvic acid dissolved in CCl$_4$ and CDCl$_3$, results of NMR analysis showed the presence of functional groups, such as aliphatic C—H, O—CH$_3$, CO$_2$CH$_3$, phenolic OH and COOH (Schnitzer and Khan, 1972). These functional groups can also be analyzed easily with infrared and other methods. Felbeck (1965) is of the opinion that the use of NMR should be limited to organic compounds that have been defined rather completely by other techniques. Ogner (1979) succeeded in quantifying the functional groups by NMR. With methylated humic acids dissolved in chloroform and ^{13}C-NMR spectroscopy, Ogner estimated that 21% of the C was aromatic, 35% in methylene chains and in methyl groups, 3% as carbonyl-C of methyl esters, and 13% of the C as methoxyl-C of methylated phenols and alcohols. The remaining 28% was not accounted for; it was suggested to be mostly C bonded to O as polysaccharides or peptides. Ogner also indicated that the ^{13}C-NMR spectrum of methylated humic acid was different from those of lignin and its derivatives.

More recently, Ruggiero et al. (1980) showed evidence for the occurrence of exchangeable aromatic protons in fulvic and humic acids in their analysis using ^{1}H-NMR spectroscopy. According to these authors the protons in the aromatic structure of humic substances could be exchanged with deuterium. The ^{1}H-NMR spectra enabled the proportion of exchangeable protons, expressed as percentage of total aromatic protons, to be measured. The estimated exchangeable proton concentrations were 18 ± 7% in humic acid and 35 ± 10% in fulvic acid.

Currently, NMR spectroscopy is considered an essential tool in the analysis of humic matter. It is capable in differentiating different types of humic matter or humic matter from different origins. The disadvantage of NMR spectroscopy is that the instruments are very expensive, and one analysis may take 24-hr scanning time or more. Nuclear magnetic resonance analysis can be performed with both liquid or solid samples of humic matter. With the very powerful instruments available today, cross-polarization–magic angle spinning carbon 13 NMR, or better known as CP–MAS ^{13}C-NMR, spectroscopy produces better spectra with solid than with liquid samples of humic acids. Soil samples can also be used in the undisturbed state, provided the sample contains sufficient amounts of organic matter. As shown in Figure 4.13, ground lignite yields an NMR spectrum with signals exhibiting an excellent resolution. The NMR spectrum of humic acid can be divided into several regions (Hatcher et al., 1980):

1. Aliphatic C region, at 0–105 ppm chemical shift: A polysaccharide is often distinguished from 65 to 105 ppm.
2. Aromatic C region, at 105–165 ppm chemical shift.
3. Carboxyl C region, at 165–185 ppm chemical shift.

The intensity of the signals for aliphatic, aromatic, and carboxyl C may vary considerably, depending on origin (differences in soils) and

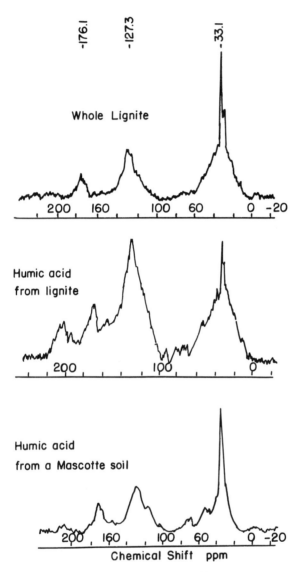

Figure 4.13 Solid state CP–MAS ^{13}C-NMR spectra of whole lignite, humic acid extracted by the NaOH method from lignite, and a Mascotte soil (sandy, siliceous, Thermic Ultic Haploquod), respectively. (From Tan, K.H., personal files.)

type of humic matter analyzed. Fulvic acid exhibits a spectrum dominated by the signals for aliphatic C, whereas the signals for aromatic C are often manifested as very weak signals. The composition of fulvic acid from aquatic humic matter is dominated especially by the aliphatic groups. This resolution forms the basis for calculating the group composition in percentages by integration of the spectrum. The data in Table 4.6 show that lignite is more aliphatic than its extracted humic acid fraction, which is high in aromatics. In general it is noted that geologic humic acid is more aromatic than aquatic or terrestrial humic acids.

In addition to the foregoing, NMR spectra can also serve as fingerprints of humic compounds. The present author used NMR (Figure 4.14) to study adsorption of fulvic acids by soils. Fulvic acid, extracted from poultry litter, exhibits an NMR spectrum composed of at least four peaks (see top curve; Figure 4.14). After shaking the fulvic acid solution with soil, the remaining supernatant was sampled and analyzed again. The NMR spectrum is shown at the bottom part of the figure. It can be noticed that peaks 1, 2, and 3 have decreased sharply in intensity, and peak 4 has disappeared. The latter suggests that the fulvic acid concentration in solution has been decreased, because of adsorption by soil.

Potentiometric Titrations of Humic Acids

Potentiometric titration has been used extensively to characterize humic substances (Posner, 1964). The curves are usually presented as sigmoidal, whereas titration was carried out mostly with a base.

Table 4.6 Aliphatic, Aromatic, and Carboxyl C Content of Lignite, Its Humic Acid Fraction, and Humic Acid Extracted From a Mascotte Soil (Spodosol)

Materials	Aliphatic C (%)	Aromatic C (%)	Carboxyl C (%)	Aromaticity (%)	Aliphaticity (%)
Lignite	63.3	30.6	6.1	32.6	67.4
Humic acid from lignite	27.2	63.8	9.0	70.1	29.9
Humic acid from soil	37.3	51.6	11.1	58.0	42.0

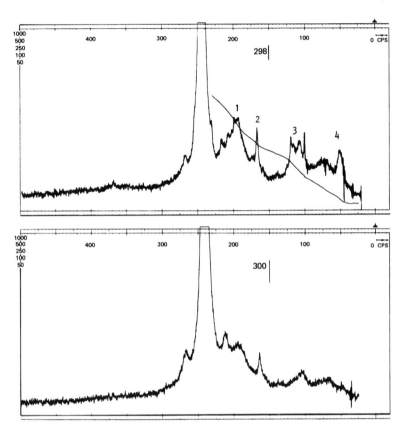

Figure 4.14 Nuclear magnetic resonance (NMR) analysis of fulvic acid extracted from chicken manure, before (top) and after (bottom) adsorption by Cecil (Ultisols) surface soil.

This result leads to the belief that humic substances have a monobasic character (Schnitzer and Khan, 1972). Humic acids are, in fact, amphoteric and polybasic. Depending on the soil condition, they can be neutral, negatively charged, or positively charged. The negative charges are attributed to the presence of phenolic-OH and carboxyl groups. The dissociation of H^+ ions from these functional groups for the development of the negative charges can be illustrated with a carboxyl group as follows:

$$R\text{—}COOH \rightleftharpoons R\text{—}COO^- + H^+$$

for which the pK value according to the Henderson–Hasselbalch equation is

$$pK_1 = pH - \log \frac{(COO^-)}{(COOH)}$$

For pK values in the acid and basic range, and the corresponding titration curves, reference will be made later.

Not only do humic acids possess two different types of oxygen-containing functional groups, they also contain dissociable H^+ ions from amino groups. Although several investigators believe that humic substances should be free of N (Burges, 1960), elemental analysis revealing significant amounts of N (see Table 4.1) suggests that N-containing compounds, such as amino acids, must be present in the humic molecule. Because amino acids are amphoteric and, depending on conditions, can behave as an acid or a base, the latter necessitates continuation of titration in the acid range. Titration curves obtained by acid and base titration of humic acids are shown in Figure 4.15a,b. The curves indicate that at pH 7.0 the carboxyl groups of humic acids are unprotonated, whereas the OH or amino groups, or both, appear to be protonated. The latter may produce substituted hydronium ions or substituted amino ions $(-NH_2-H)^+$. Addition of acid to the solution lowers the pH rapidly at first and then more slowly as the buffering action of the carboxyl groups is activated. At pH 2.4 the pK_a is reached, at which half of the carboxyl groups of humic acid (high molecular weight) are considered ionized, conforming to the Henderson–Hasselbalch equation. Further addition of acid results in a slight decrease in pH, and finally, at pH 2.0, the high-molecular-weight humic acid starts to flocculate. Titration of the substituted OH and amino acids with base follows an opposite trend in the alkaline region. According to the inflection point, the pK_a is reached at pH 11.2.

Chromatography of Humic Substances

Two types of chromatographic methods have been used in the study of humic material—gel chromatography and gas–liquid chromatography—for quite different purposes. Gel chromatography is proposed for purification processes or for fractionation of the humic compounds into different components or into fractions of different molecular sizes. The latter has been extended to measurements of molecular weights (Orlov et al., 1975; Holty and Heilman, 1971;

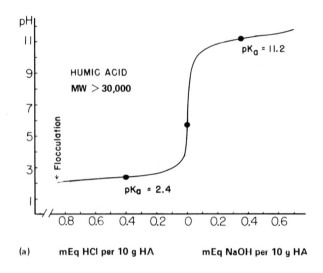

(a)   mEq HCl per 10 g HA   mEq NaOH per 10 g HA

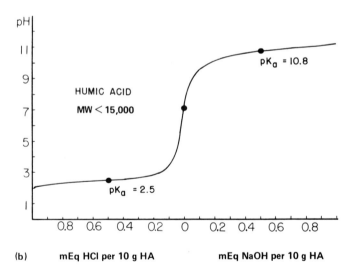

(b)   mEq HCl per 10 g HA   mEq NaOH per 10 g HA

Figure 4.15 Titration curves of (a) high (>30,000) and (b) low (<15,000) molecular weight fractions of humic acid from a Cecil soil (Ultisols). Molecular fractionation was done by gel chromatography using Sephadex G-50. (From Tan and Giddens, 1972.)

Swift et al., 1970; Mehta et al., 1963). On the other hand, gas–liquid chromatography is used primarily to fractionate the humic material into its component molecules. However, both methods can also be applied in the characterization of humic substances by using the chromatograms as fingerprints. Unfortunately, the advantages of the latter have not been fully exploited. Today the use of gas chromatography together with mass spectrometry is considered a powerful tool for structural analysis of humic acids (Schnitzer, 1976).

Gel Chromatography

Gel chromatography is a simple and relatively effective method in achieving molecular fractionation. A variety of materials have been used as the gel substance (e.g., polysaccharides, polystyrene, polyamides, aluminum oxides, cellulose, agar, and glass beads). The

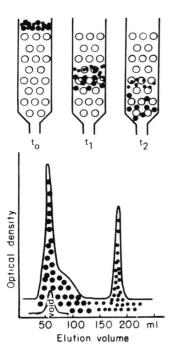

Figure 4.16 Sephadex G-50 gel filtration of humic acid from a Cecil soil (Ultisols): (●) high molecular weight HA, (•) low molecular weight HA, (○) Sephadex beads.

most widely used gel materials are the cross-linked polymers of polysaccharides, polystyrene, polyamides, and the like, distributed under names such as Sephadex, Sephagel, Bio-gel, Cellogel.

A column of swollen gel beads is prepared, and a solution of humic acid is eluted (filtered) through the column at a controlled flow rate. The filtration is diagramatically illustrated in Figure 4.16. The pores between and within the gel beads enable the gel column to act as a chromatographic medium. The large open circles in the figure are the Sephadex beads. The small and large black dots represent a mixture of small and large molecules of humic acids. As the mixture of molecules passes through the column, the larger molecules are eluted first and, depending on conditions, the elution curve may be represented by two peaks, as shown in the figure. The first peak is attributable to the larger molecules and the second peak to the smaller molecules. If the mixture is composed of only small molecules, such as fulvic acids, the elution curve may be characterized by only one major peak (Figure 4.17).

The behavior of the solute in gel filtration can be measured by its elution volume, whereas the elution peak is determined by the partion coefficients K_{av} and K_d, which are defined as

$$K_{av} = \frac{V_e - V_0}{V_t - V_0} \qquad K_d = \frac{V_e - V_0}{V_i}$$

where

V_e = elution volume
V_0 = void volume
V_t = total volume of gel bed
V_i = inner volume of the gel
        (*Handbook Sephadex-Gel Filtration in Theory and Practice*, Pharmacia Fine Chemicals, Uppsala, Sweden, 1969).

Tan and Giddens (1972) reported that the partition coefficients K_d were higher for the small molecules of humic material, and a positive correlation was obtained between K_d values of humic fractions separated with Sephadex and color ratios (E_4/E_6 ratios). As can be noticed from Figure 4.18, humic fractions with higher partition coefficients exhibit larger E_4/E_6 ratios. This corresponds to the observation indicated that the smaller molecules, eluted last, also

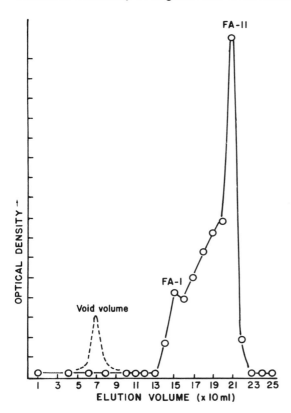

Figure 4.17   Gel filtration of fulvic acid using Sephadex G-50. FA-I has larger-sized molecules than FA-II.

have lower molecular weights, whereas the larger molecules, eluted first, are of higher molecular weights.

Gas–Liquid Chromatography

Gas–liquid chromatography (GLC) is a complex method that is not as versatile as gel chromatography. Gas chromatography requires that the material to be analyzed can easily be transferred into the gaseous phase, which is a major obstacle with humic acids. Humic compounds are generally nonvolatile, and no method has been yet reported in the literature for making them volatile without decomposing the humic material into smaller components first. Schnitzer

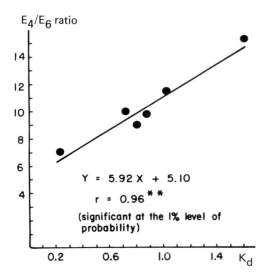

Figure 4.18 The relationship of color ratios (E_4/E_6 ratios) of fulvic acids and partition coefficients (K_d values). (From Tan and Giddens, 1972.)

and Khan (1972) reported the senior author's efforts in gas chromatography using degradation products of fulvic acids, obtained by oxidation with strong acids or by reduction with sodium amalgam. The degradation products identified were mostly benzene derivatives (e.g., benzene di-, tri-, tetra-, penta-, and hexacarboxylic acids). Although valuable information can thus be obtained, much remains unknown concerning the natural assembly patterns in humic acids. The extent to which organic artifacts that no longer resemble the real structural units of the original humic material may be produced during the oxidation or reduction processes is unclear.

Success has been obtained in gas chromatography of carbohydrates and related polyhydroxy compounds with the use of trimethylsilyl (TMS) (Tan and McCreery, 1970a). The TMS group $Si(CH_3)_3$, was then introduced (by shaking) into humic substances through their functional groups, which have exposed protons or active H^+ ions:

$$R—COOH + TMS—Cl \rightarrow R—COO—TMS + HCl$$

Silylated humic acid from different kinds of soil exhibited gas chromatograms (Figure 4.19) with almost the same number of compo-

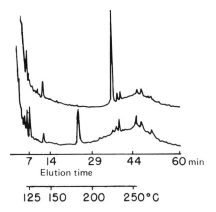

Figure 4.19   Gas chromatograms of humic acids silylated for 36 hr. Methylated HA from a Cecil soil (Ultisols, top); methylated HA from a Greenville soil (Ultisols, bottom). (From Tan and McCreery, 1970b.)

nents appearing at similar retention times (Tan and McCreery, 1970b). However, the concentration varied considerably between some of the components of the two humic acids, as can be seen by the differences in peak height.

Molecular Weights of Humic Acids

The use of molecular weights in characterization of humic substances encounters many problems because these compounds are known to be polydispersive. They possess, therefore, a wide spread in molecular weights, causing the humic material to be very heterogeneous in this respect (Felbeck, 1965). The degree of polydispersity may vary considerably depending on the contributing components of the humic molecule with different molecular weights.

Physically, molecular weights can be expressed as follows:

1. The *number–average molecular weight* $\overline{M}_n$ is defined as

$$\overline{M}_n = \frac{\sum nM}{n}$$

where

n = number of component molecules
M = molecular weight of component molecules

The methods used to determine $\overline{M}_n$ are osmometry, diffusion, and isothermal and cryoscopic distillation. Osmometry is considered the best method, but it appears not to be applicable to analysis of molecular weights >200,000.

2. The *weight–average molecular weight $\overline{M}_w$* is defined as

$$\overline{M}_w = \frac{\sum nM^2}{\sum nM}$$

The weight–average molecular weight is usually measured using viscosity analysis and gel filtration. Of the two, gel filtration is the simplest method, as discussed earlier.

3. The *Z–average molecular weight $\overline{M}_z$* which is defined as:

$$\overline{M}_z = \frac{\sum nM^3}{\sum nM^2}$$

The method used to measure $\overline{M}_z$ is the sedimentation method employing the ultracentrifuge. It yields many problems with humic compounds. Humic acid carries a negative charge balanced by cations, thereby creating a diffuse double-layer system. Because of the latter, the molecules tend to repel each other, offsetting the sedimentation process. Intermolecular repulsion yields high-diffusion and low-sedimentation coefficient values owing to faster sedimentation of the larger molecules than the counterions, resulting in an electrostatic drag. In addition, the polydisperse nature makes it difficult to achieve well-defined sediment boundaries with humic material.

For a heterogeneous, or polydisperse, system, $\overline{M}_n < \overline{M}_w < \overline{M}_z$, but for a homogeneous, or monodisperse, system, $\overline{M}_n = \overline{M}_w = \overline{M}_z$.

Values reported for weight–average molecular weight M_w of humic substances may vary from 1000 to 30,000. Flaig and Beutelspacher (1951) stated molecular weights of >100,000, and values of 2 million have been reported occasionally. Apparently, any number within these ranges can be obtained, depending on the chemical isolation procedures and analysis employed, with fulvic acids usually exhibiting the lower and humic acids the higher-molecular-weight values. Tan and Giddens (1972) reported that with Sephadex G-50 gel filtration, humic acid was separated only into two fractions:

a high (>30,000) and a low (<15,000) molecular weight fraction (Figure 4.20). The high-molecular-weight fraction made up 50% of the humic acid isolate, whereas the remaining 50% was the low-molecular-weight humic compound. With the same gel filtration technique, Tan and McCreery (1975) noted that the degree of po-

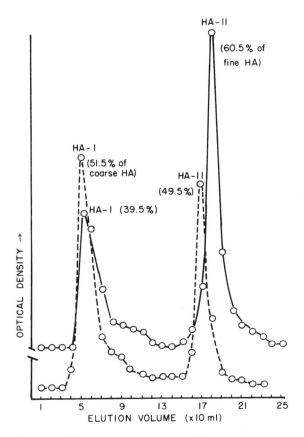

Figure 4.20 Sephadex G-50 gel filtration of humic acids. HA-I, high-molecular-weight fraction of humic acid; HA-II, low-molecular-weight fraction of humic acid. Coarse HA (broken line) represents dialysis of humic acid conducted with a membrane with coarse pores (molecular weight cutoff of membrane is 10,000). Fine HA (solid line) represents dialysis of humic acid conducted with a cellulose membrane with fine pores (molecular weight cutoff of the membrane is 3500).

Table 4.7 Molecular Weights and Size (in Å and nm) of Humic Fractions Obtained by Sephadex Gel Filtration

Molecular weight	Molecular volume (Å³)	Radius (Å)	(nm)
30,000	23,622	17.8	1.78
5,000	3,937	9.8	0.98
1,500	1,181	6.6	0.66
1,000	787	5.7	0.57

lymerization or the size of the molecules isolated affected molecular weight. The data in Table 4.7 demonstrate the relation between the size of molecule and molecular weight. By assuming that the humic molecules are spherical, the larger the size of the molecules of the humic fraction isolated, the larger will be the numerical value of the molecular weight of humic acid.

Electron Microscopy of Humic Acids

Flaig and Beutelspacher (1951) were perhaps among the first who tried electron microscopy in the study of humic acids. This method, which enjoyed considerable success in clay mineralogy, was used to analyze the morphology and dimension of humic particles. From transmission electron microscopy, Flaig and Beutelspacher indicated that humic acids occurred as very small spherical particles, of about 100–150 Å (10–15 nm) in diameter. The spheres frequently joined together in racemic chains. A more recent report on transmission electron microscopy (TEM) studies of HA and FA indicated that, in dilute aqueous solutions, humic acids form small spheroids, 9–12 nm in diameter, connected by long multibranched fibers, 20–100 nm in width (Stevenson and Schnitzer, 1982).

Scanning electron microscopy (SEM) has also become an important tool for the investigation of shape, size, and degree of aggregation of fulvic and humic acids and their complexes (Chen and Schnitzer, 1976; Dormaar, 1974). This method has an advantage over standard TEM by furnishing a three-dimensional picture with a resolution depth of 5–10 μm. In addition, humic particle surfaces and orientation can also be observed. Several methods have been proposed for sample preparation to avoid altering or damaging the original structure of humic molecules. Chen and Schnitzer (1976) used a freon–liquid N_2-freezing technique for their SEM analysis. Humic solutions were first frozen on mica sheets or glass slides and then transferred to a sample holder for scanning. Their SEM micrographs (Figure 4.21) showed that protonated fulvic acid (adjusted to pH 2–3) exhibited an open structure, formed from elongated fulvic acid strands or fibers with rounded or spherical tips. The structure changed into a spongelike structure if the pH of fulvic acid was adjusted to 7.0, whereas at pH 9.0 fulvic acid revealed homogenous sheets, in which grains were visible. The behavior of macromolecular structures of humic acids appeared to vary not only according to pH, but also according to ionic strength and sample concentration (Ghosh and Schnitzer, 1980c). At high concentration, low pH, or at medium pH with high ionic strength, they behaved as rigid spher-

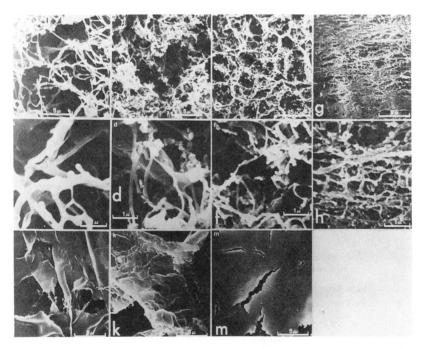

Figure 4.21 Scanning electron micrographs·of fulvic acids at various pH values: (a),(b) at pH 2.0, (c),(d) at pH 4.0, (e),(f) at pH 6.0, (g),(h) at pH 7.0, (i) at pH 8.0, (k) at pH 9.0, (m) at pH 10.0. (From Chen and Schnitzer, 1976.)

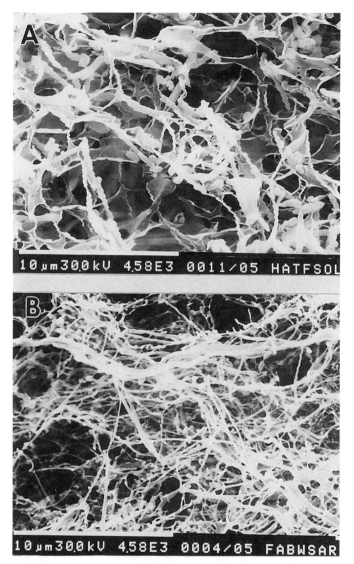

Figure 4.22 Scanning electron micrographs of humic and fulvic acids from specimens prepared by the revised liquid N_2 freeze-drying method. (a) Humic acid extracted by the NaOH method from a Tifton soil; (b) fulvic acid isolated by the XAD-8 resin method from black water of the Satilla river in Georgia.

ocolloids, but were otherwise flexible linear colloids. A revised procedure for SEM preparation of samples by rapid freeze-drying in liquid N_2 was introduced by Tan (1985). To make the procedure simpler and faster, the use of freon gas and prior preparation on glass slides or mica strips were deleted. The liquid HA sample was immediately placed on a SEM sample stub (holder) and frozen quickly in liquid N_2. The frozen sample was dried in a vacuum evaporator at room temperature while a high vacuum (6.5×10^{-10} MPa) was maintained for 12 hr. This method has been adapted from preparation of animal tissue for surface scanning electron microscopy. The results of the modified rapid liquid N_2 technique were similar to those of the freon–liquid N_2 method of Chen and Schnitzer (Figure 4.22).

Structural Chemistry of Humic Acids

Several hypotheses have been reported on the structural chemistry of humic acids, but apparently the hypotheses lack desirable uniformity and much disagreement still exists.

Hypothesis of Schnitzer and Khan (1972). The concept is based on information obtained from chemical degradation of fulvic acids. Degradation reactions do not ensure that artifacts may not have been produced. Depending on the severeness or mildness of the reactions applied, any breakdown product can be obtained, ranging from elemental C, H, O, benzene rings to heterocyclic rings. Schnitzer and Khan are of the opinion that humic substances must be broken down into smaller subunits to study their structural chemistry. Thus far four basic types of degradation procedures have been used:

1. Oxidation with alkaline permanganate, nitric acid, H_2O_2, and CuO–NaOH mixture. The degradation products were invariably benzenecarboxylic acids.

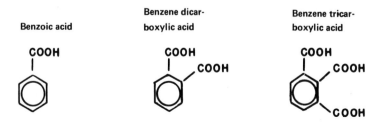

Benzoic acid   Benzene dicarboxylic acid   Benzene tricarboxylic acid

COOH

,COOH

`COOH

COOH

Benzene tetracar-
boxylic acid

COOH

,COOH

HOOC`     `COOH

COOH

Benzene pentacar-
boxylic acid

2. Reduction with Na amalgam or with Zn dust. Fulvic acid also yielded benzene derivatives with this method.
3. Hydrolysis with hot water, with acids or bases. Fulvic acid yielded benzene derivatives such, as hydroxybenzoic and vanillic acids.
4. Biological degradation: This is a method by which fulvic acid is decomposed with the aid of microorganisms (e.g., *Penicillum* sp., *Aspergillus* sp., *Trichodermia* sp.). The compounds produced from fulvic acid by biological degradation were also identified as benzene derivatives.

On the basis of the predominant findings of benzene derivatives, Schnitzer and Khan (1972) assume that fulvic acid is composed of phenolic and benzene carboxylic acids, joined together by hydrogen bonds to form a polymeric structure. The latter contains many voids or openings in which other organic compounds, such as amino acids and carbohydrates, can be trapped.

Hypothesis of Kononova (1961). Kononova is of the opinion that at least three basic steps are involved in formation of humic acids: formation of structural units from the decomposition of plant tissues, condensation of these units, and polymerization of the condensation products. The result is a multicomponent system, called humic or fulvic acids. They show similar structural patterns but may differ in details of structural and chemical composition; for example, fulvic acid has a less condensed aromatic nucleus, but has more highly developed peripheral components. In Kononova's opinion fulvic acid can be both the predecessor and the decomposition product of humic acid.

The basic structural units of humic compounds are considered to be phenolic or quinoid, bonded to nitrogen-containing compounds and carbohydrates, the latter chiefly polyuronides. The inclusion of

N-containing compounds and carbohydrates as fundamental units of the humic molecule is a matter of much controversy. Several investigators regard the latter as accidental contaminants trapped in the maze work of the humic structure (Burges, 1960; Schnitzer and Khan, 1972), but others show evidence for the necessary participation of carbohydrates and N compounds in the formation of humic acids (Kononova, 1961; Flaig et al., 1975). Kononova (1961) suggests that the following reaction occurs for the inclusion of N in the humic molecule:

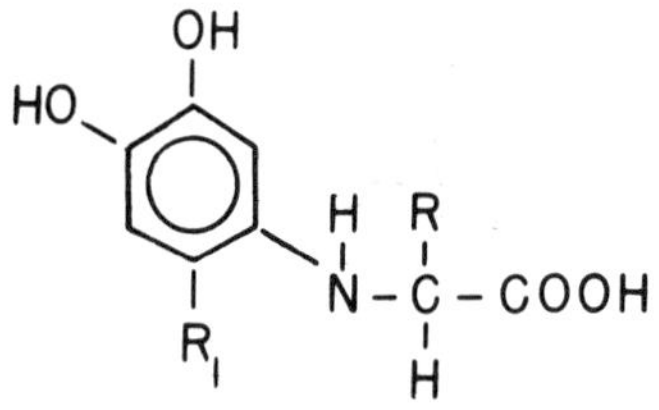

Such a combination produces a stable condensation product of phenols and α-amino acids and increases the stability of N in acid hydrolysis.

Several authors disagree on including carbohydrates in the humic molecule; they are of the opinion that they are contaminants that can be removed by adequate purification methods. Notwithstanding these arguments, various research reports (Clark and Tan, 1969; Tan and Clark, 1968; Tan, 1975) point to the discovery of polysaccharides as an integral part of the humic molecule. Fulvic acid appears to be composed of large amounts of polysaccharides, and hymatomelanic acid has been identified as a polysaccharide ester compound. Infrared spectroscopy of hymatomelanic acid shows a characteristic strong absorption between 3000 and 2800 and at 1720 cm^{-1}, the latter, usually accompanied by a weak shoulder at 1620 cm^{-1} (Figure 4.23).

These absorption features can be reproduced artificially, or closely simulated, by methylation of humic acid. In addition, the infrared absorption of hymatomelanic acid at 1620 cm^{-1} can be increased considerably in intensity by separation of the ester group through hydrolysis of the compounds with 2 N NaOH. The latter procedure removes the blocking effect of the ester group on the carboxyl vibration in COO^{-} form. Identification of the ester fraction reveals

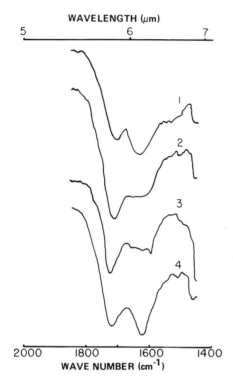

WAVELENGTH (μm)

WAVE NUMBER (cm⁻¹)

Figure 4.23 Infrared spectra of humic fraction between 2000 and 1400 cm^{-1}, showing absorption attributed to carbonyl (1720 cm^{-1}) and carboxyl (COO^{-}, 1625 cm^{-1}) groups. (1) Purified humic acid, (2) hymatomelanic acid, (3) methylated humic acid, and (4) hymatomelanic acid, after removal of ester group. (From Tan and McCreery, 1970b.)

infrared features similar to those of a soil polysaccharide (Mortenson, 1961).

Ogner (1980) indicated that polysaccharides in humic compounds were highly branched and complex in composition. They contained at least 34 different aldose building units, in addition to O-methyl, di-O-methyl, amino derivatives, ketones, and uronic acids. Generally, polysaccharides of fulvic acids were comparatively less branched than those of humic acids.

Hypothesis of Flaig. Flaig and co-workers (Flaig et al., 1975) suggest lignin to be the source, or starting point, for the formation of humic and fulvic acids. Lignin is assumed to be broken down by degradation or decomposition reactions into its basic units (i.e., coniferyl alcohol or guaiacyl propane monomers). These lignin basic units are then subject to oxidation, followed by demethylation to substituted polyphenols and further oxidation to quinone derivatives. Condensation of the quinone groups with amino acids and polysaccharides may then yield humic acidlike substances. Lignin degradation products have been detected in hydrolysates of humic acids.

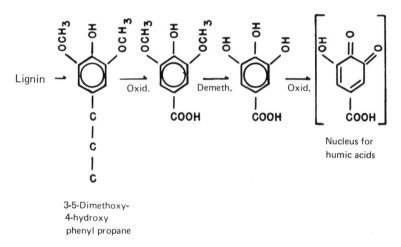

3-5-Dimethoxy-
4-hydroxy
phenyl propane

Author's Hypothesis.   Among the different opinions existing on the origin and structure of humic matter, an agreement is present indicating that phenolic groups and nitrogen compounds are the building blocks for humic matter. The latter concept is reflected in the *lignoprotein theory*, which differs only slightly from hypotheses derived from oxidation and degradation research on humic compounds. Oxidation of humic acids yields phenols and a series of phenol-carboxylic acids, which form humic acids by condensation and polymerization (Schnitzer and Khan, 1972). The general opinion is that humic acid cannot be described in definite chemical terms, but current research data point otherwise. Results of ^{13}C-NMR analyses suggest the presence of a consistent structural composition, including aliphatic, aromatic, and carboxyl compounds. Infrared spectroscopy also provides evidence for only one basic structure related to the presence of aliphatic C—H, carboxyl, and phenolic-OH functional groups. In addition, the elemental composition is fairly consistent with C content varying only from 40 to 50%, H from 4 to 5%, N from 1 to 5%, and S from 0.2 to 0.3%. For these reasons, the present author proposes the following basic structure for the smallest possible molecule of terrestrial humic acid (Figure 4.23). The aromatic nucleus can vary depending on the type of the lignin monomer. The figure also shows the possibility to add a carbohydrate

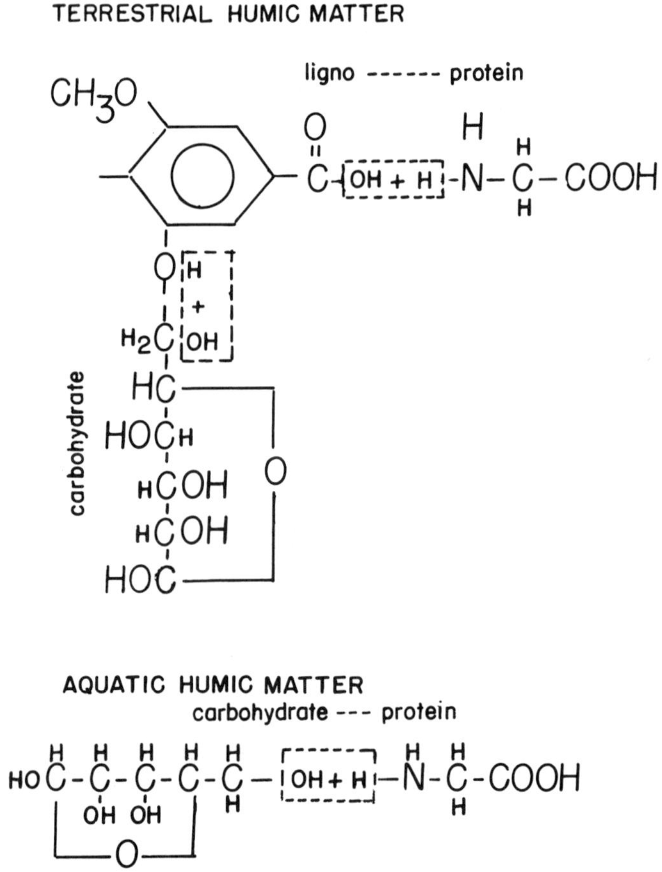

Figure 4.24 Basic structure, or "unit cell," of a humic molecule according to the lignoprotein and carbohydrate–protein theories, respectively.

molecule to the aromatic nucleus. The latter is of importance for the structure of fulvic acids, since fulvic acids contain less N and more carbohydrates, as noticed from both NMR and infrared analysis. A complex structure can also be built by combining several units of the basic molecule, or adding other compounds to the aromatic nucleus. Since the basic structure of a humic molecule, as just outlined, can repeat itself in any direction, it is, therefore, comparable to a "unit cell" in clay mineral structures.

Effect of Humic Acids on Plant Growth

The effect of organic matter on plant growth has been known for some time. The major benefits of soil humus on plant growth result indirectly through improvement of soil properties, such as aggregation, aeration, permeability, and water-holding capacity. Humic acids are also capable of decreasing aluminum toxicity in soils. Acidic soils may contain soluble and free Al in amounts toxic for plant growth. By chelating the excess Al, humic acid reduces the concentration and chemical activity of free Al to the benefit of plant growth (Tan and Binger, 1986, Ahmad and Tan, 1986). In general, adding humic acids to acid soil can take excess metal cations out of the solution and store them for later use by plants. Thus, humic acid prevents the buildup of large amounts of micronutrients in the soil and, at the same time, releases them back in amounts suitable for plant growth. With the increased knowledge of soil organic matter and, in particular, of humic acid chemistry starting a decade ago, increasing amounts of information have been accumulated on the direct influence of soil organic matter and humic acids on plant growth. Unconfirmed reports also mention that nucleic acids, protein, and small degradation products of humic acids can be taken up by plants (Flaig, 1975). Organic compounds, such as acetic, propionic, butyric, and valeric acids, have been noted to increase root growth only when they were present in combination. Alone, they had no effect, as studied by Wallace and Whitehand (1980) with root elongation of germinating wheat on agar media. In summary, it can be stated that humic acid compounds and the like can improve plant growth directly by acceleration of respiratory processes, by increasing cell permeability, or by hormone growth action. Most of the investigations with humic acids, limited to studying seed germination, shoot growth, and elongation of very young seedlings, or root

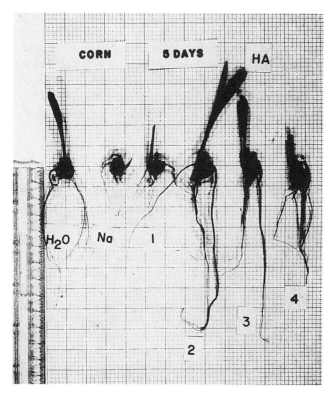

Figure 4.25 Germination and growth of 5-day-old corn seedlings as affected by humic acid: (H_2O) 0 ppm HA, (Na) blank + 0.66 mEq NaOH, (1) 320 ppm HA, (2) 640 ppm HA, (3) 1600 ppm HA, (4) 3200 ppm HA. The corn was grown in a modified Hoagland solution to which the HA treatments were applied. (From Tan, 1978; Tan and Nopamornbodi, 1979.)

elongation of excised roots in vitro, show the presence of a hormonal growth effect (Poapst et al., 1970, 1971). On the other hand, work done on nutrient uptake by Guminski and others in East Europe (Guminski et al., 1977; Guminski, 1957) and by Dormaar (1975) in Canada reveals the physiological influence of humic acids on plant growth. From studies on the growth and nutrient uptake of corn plants (*Zea mays* L.), Tan and Nopamornbodi (1979) came to the conclusion that humic acid affected plant growth through a combination of the aforementioned processes. As can be noticed from

Figure 4.25, moderate amounts of humic acid were generally beneficial to root and shoot growth of corn plants. At the same time, a significant increase in N content of shoots of corn seedlings was obtained. Dry matter production also appeared to be stimulated by moderate amounts of humic acids.

5

Colloidal Chemistry of Inorganic Soil Constituents

5.1 THE CLAY FRACTION OF SOILS

The inorganic fraction of soils is composed of rock fragments and minerals of varying size and composition. Despite the variability in composition, the inorganic fractions are predominantly silicates and oxides. They are sometimes distinguished into primary and secondary minerals. *Primary minerals* are, by definition, rock-forming minerals that are present in soils chemically unchanged, whereas *secondary minerals* are minerals that have been formed by weathering of primary minerals. But this distinction creates problems, since secondary minerals may well be regarded as primary on a pedological basis. On the basis of size, three major fractions are usually recognized: (1) the coarse fraction (2–0.050 mm) called *sand*, (2) the fine fraction (0.050–0.002 mm) called *silt*, and (3) the very fine fraction (<0.002 mm) referred to as *clay* (USDA, 1975). In soil science, we are used to considering clay as a colloid, although strictly speaking only the fine clay fraction <0.2 μm is colloidal clay.

Six types of soil silicates are usually recognized on the basis of the arrangement of the SiO_4 tetrahedra in their structure:

1. *Cyclosilicates*: Closed rings or double rings of tetrahedra (SiO_3, Si_2O_5)
2. *Inosilicates*: Single or double chains of tetrahedra (SiO_3, Si_4O_{11})
3. *Nesosilicates*: Separate SiO_4 tetrahedra
4. *Phyllosilicates*: Sheets of tetrahedra (Si_2O_5)
5. *Sorosilicates*: Two or more linked tetrahedra (Si_2O_7, Si_5O_{16})
6. *Tectosilicates*: Framework of tetrahedra (SiO_2)

Examples of mineral species belonging to the respective categories are listed in Table 5.1. The sand and a major part of the silt fraction are cyclo-, ino-, neso- soro-, or tectosilicates. They make up the framework of the soil. Since they are coarse in size, they have low specific surface area and do not exhibit colloidal properties. Although not really active in chemical reactions, they participate in a number of reactions and exhibit some adsorption. Many of the sand and the silt minerals are also of importance for formation of clays. Some of the clays are phyllosilicates.

Clays are of special importance in soil chemistry, since they have a surface chemistry different from that of the larger mineral grains. Clays also exhibit bulk physical properties different from gravel, sand, or silt. Many of the minerals in soil clays are crystalline in structure, whereas others may poorly exhibit crystals or be structurally disordered. Some of the clays may be amorphous (e.g., silica, alumina, and iron oxide gels). The latter may occur in soils as discrete, independent minerals, or as coatings around crystalline clay particles and other inorganic soil constituents. Not all of the clays belong to the phyllosilicate group, or layer–lattice silicates. The

Table 5.1 Examples of Mineral Species Classified According to the Six Types of Soil Silicate

Soil silicate	Mineral species
Cyclosilicates	Tourmaline
Inosilicates	Amphibole, pyroxene, hornblends
Nesosilicates	Garnet, olivine, zircon, topaz
Phyllosilicates	Chlorite, illite, kaolinite, montmorillonite, vermiculite
Sorosilicates	Epidote
Tectosilicates	Feldspars, zeolite

Table 5.2 Major Phyllosilicate Minerals in Soils

Layer type	Group name	Charge per unit formula	Common minerals
1:1	Kaolinite–serpentine	~0	Kaolinite, halloysite Chrysotile, lizardite, antigorite
2:1	Pyrophyllite–talc	~0	Pyrophyllite and talc
	Smectite, or montmorillonite– saponite	0.25–0.6	Montmorillonite, beidellite, nontronite Saponite, hectorite, sauconite
	Mica	~1	Muscovite, paragonite Biotite, phlogopite
	Brittle mica	~2	Margarite, clintonite
	Illite	2	Illite
	Vermiculite	0.6–1.9	Vermiculite
2:1:1	Chlorite	variable	Chlorite
Chain	Palygorskite– sepiolite	—	Palygorskite, sepiolite

Source: Mackenzie (1975) and Brindley et al. (1968).

soil's clay fraction also contains other minerals such as the paly-gorskite–sepiolite minerals, which are chain-structured minerals, quartz in particle sizes of <2 μm; sesquioxides, titanium oxides, pyrophyllite, talc, sulfides, sulfates, and phosphates. The major types of phyllosilicates are listed in Table 5.2.

5.2 STRUCTURAL CHEMISTRY OF CLAY MINERALS

As indicated earlier, soil clays can exist in crystalline, structurally disordered, or amorphous form. The amorphous state generally has no recognizable shape or geometric internal arrangement of atoms. Depending on the degree of sophistication in methods for analysis, a sharp distinction does not exist between crystalline and amorphous states. In soil science, clay is considered amorphous if it is amorphous to x-ray diffraction analysis (i.e., lacks regularity in in-

ternal atomic arrangement as reflected by a featureless diffracto-
gram). In crystals, the atomic arrangement may be repeated in a
regular three-dimensional pattern. However, in amorphous materials, such as glass, the chemical bonding of the component atoms
is, perhaps, the only unit repeating itself. The spatial arrangement
of atoms producing the building unit of a crystal is called the *unit
cell*. The latter exhibits a complete group pattern of atoms that repeats itself in three directions in space according to the so called x,
y, and z axes, respectively (Figure 5.1). The z axis is sometimes called
the c axis. The size or length of the edges of the unit cell in each
direction is expressed in terms of a, b, and c, each of which represents

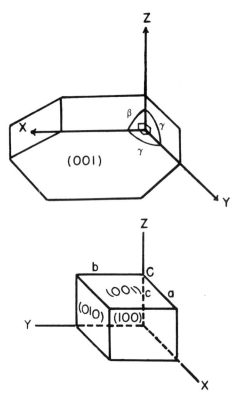

Figure 5.1 (Top) Crystal structure, exhibiting a group pattern of atoms,
repeating itself in three directions in space, according to the so-called x, y,
and z axes, respectively. (Bottom) Close-up of a unit cell, showing discrete
unit length a, b, and c, measured along the x, y, and z axes in a cubic crystal.

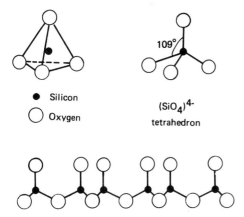

Figure 5.2 (Top) Schematic structure of a single silica tetrahedron. (Bottom) The arrangement of several silica tetrahedra into a sheet by mutually sharing oxygen atoms.

a discrete unit length for a specific crystal, measured along the x, y, and z axes. In a cubic crystal, a, b, and c are of equal length, and the angles, α, β, and γ are 90°. In clay minerals, these angles may vary according to the structure. By placing several unit cells together, the crystal arrangement produced is then called a *lattice structure*. A perfect crystal may be composed of unit cells, each of which has a volume of approximately 1 μm^3.

The atomic groups in a crystal lattice can be arranged in planes at equal spacings along the crystallographic direction. Several types of atomic planes can be drawn in the crystal with interplanar spacings called *d spacings*. The plane delineated, or bordered, by a and b parallel to the x and y axes (see Figure 5.1) cuts the z axis at C, but does not cut the x and y axes. According to the Miller indices system (Grimshaw, 1971), this plane is given the number 001. The basal (001) spacing plays a fundamental role in identification of clay mineral species by x-ray diffraction analysis.

Silicates are built around a silica tetrahedron, in which each oxygen atom receives one valency from the silicon atom. To satisfy its divalent requirement, the oxygen atoms can be linked to other cations or to a silicon atom of an adjacent silica tetrahedron (Figure 5.2). The linkage of silica tetrahedra yields five groups of structural arrangements of silicates: island, isolated group, chain, sheet, and framework structure. Silicate clay minerals are characterized by a sheet structure. In contrast with the other silicates, the structure of clay is not a three-dimensional network of simple linkages of silicon–oxygen units, but it is built up of stacked layers of silica tetrahedra and Al (Mg) octahedra sheets. The sheets are developed by the linkage of three oxygens in each tetrahedron with adjacent silica tetrahedra units, as discussed earlier. The silica tetrahedra are arranged in hexagonal rings

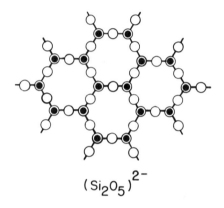

$$(Si_2O_5)^{2-}$$

and the sheet can extend indefinitely in a two-dimensional direction, according to the two planes a and b, or parallel to the plane of the paper in this book, which is the reason for the platelike nature of clays. The composition of each ring, or the lowest unit of the sheet is

$$(Si_2O_5)^{-2} = \frac{6\ Si}{3} + \frac{6\ O}{2} + \frac{(6\ O^-)}{3}$$

In such a network of silica tetrahedra, one oxygen in each tetrahedron remains electrically unbalanced. To satisfy the divalent requirement, the latter is linked to Al in octahedral coordination. By such a packing of silica tetrahedron and aluminum octahedron sheets, a layered clay structure is formed. Several layers of silica tetrahedron and aluminum octahedron sheets can be stacked one above another. However, each layer is an independent unit and is considered the crystal unit. The bonds between the layers can be relatively strong, as in kaolinite, or can be relatively weak, as in montmorillonite. Within each layer a certain atomic grouping repeats itself in the lateral direction. This group or unit layer is referred to as the *unit cell*, whereas the total assembly of a layer plus interlayer material is called a *unit structure*. Illustrations of the packing of silica tetrahedra and aluminum octahedra will be given in the sections on the major types of clay minerals.

On the basis of the number of tetrahedral to octahedral sheets in one layer, the following structural types are recognized: $1:1$ or dimorphic, $2:1$ or trimorphic, $2:2$ or $2:1:1$ or tetramorphic types. The kaolinite group represents $1:1$ layer structures, because of its composition of one tetrahedral to one octahedral sheet. The montmorillonite group represents the $2:1$ type, since its structure is built of two tetrahedral sheets to one octahedral sheet. The chlorite group is an example of a $2:2$ type, while palygorskite and sepiolite belong to the $2:1:1$ type.

Each clay mineral group can be subdivided again into two subgroups: dioctahedral and trioctahedral. If two of three of the octahedral positions are occupied by Al^{3+}, for example, it is dioctahedral. If all octahedral positions are occupied by Mg^{2+}, it is a trioctahedral subgroup.

In addition to the structural arrangement as just discussed, stacking of the layers can also occur by different types of unit layers in a regular or irregular pattern. The latter yields the interstratified group of clay minerals. The structure of these minerals may vary widely, since two or more different types of unit layers may be stacked together (e.g., vermiculite with chlorite units, chlorite with smectite units, mica with smectite units, and kaolinite with smectite units).

Kaolinite Group (1:1 Layer Clays)

Kaolinite minerals are hydrated aluminosilicates, with a general chemical composition $Al_2O_3:SiO_2:H_2O$ = $1:2:2$, or $2SiO_2 \cdot Al_2O_3 \cdot 2H_2O$ per unit cell. Structurally they are $1:1$ type phyllosilicates. The crystal is composed of aluminum octahedra sheets stacked above silica tetrahedron sheets (Figure 5.3). The sheets extend continuously in the a and b directions and are stacked one above the other in the z or c direction. The unit cell is nonsymmetric with a silica tetrahedra sheet on one side and an aluminum octahedra

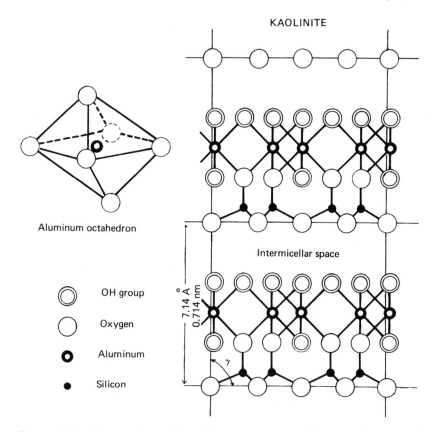

Figure 5.3 The structure of kaolinite, composed of silica tetrahedron and aluminum octahedron sheets looking down the (b) direction. Unit cell formula: $[Al_2(OH)_4(Si_2O_3)]_2$, $a = 5.14$ Å, $b = 8.93$ Å, $c = 7.37$ Å, $\alpha = 91.8°$, $\beta = 104.5°$, $\gamma = 90°$.

sheet on the other. Consequently, the basal plane of oxygen atoms in one crystal unit is opposite the basal plane composed of OH ions of the next layer. The latter gives the mineral two types of surfaces. The two sheets forming a unit layer are held together by oxygen atoms that are mutually shared by the silicon and aluminum atoms in the respective sheets. The unit layers, in turn, are held together by hydrogen bonding, yielding an intermicellar space with fixed dimension. The basal (001) spacing of kaolinite is therefore 7.14 Å.

There is little isomorphous substitution, and the permanent charge per unit cell, if not zero, is very small. However, owing to the presence of exposed hydroxyl groups, kaolinite has a variable, or pH-dependent, negative charge. As can be noticed from its structure, the position of the OH group opens possibilities for dissociation of H^+, which is the reason for the development of the variable charges, especially at the plane of hydroxyl groups on the exposed surface of the octahedral site. Another plane of hydroxyls is also present, but the latter is located as a subsurface plane of the octahedrons, covered by a network of oxygen ions. The possibility for dissociation of H^+ ions through such a network of oxygens is still unknown. The cation exchange capacity of kaolinite is, therefore, very small and may change with pH. Usually, it is in the range of 1–10 mEq/100 g.

Because of the tightness of the structural bonds, kaolinite particles are not easily broken down. This is also the cause for low plasticity and shrinkage and swelling properties. Its restricted surface limits the adsorption capacity for cations. The specific surface area is approximately 7–30 m^2/g.

The presence of kaolinite can be identified by a (001) d spacing of 7.14 Å x-ray diffraction, and by a second-order 3.57 Å diffraction in oriented samples. These diffraction peaks will disappear after heating kaolinite to 500–550°C.

Members of the kaolinite group are kaolinite, dickite, nackrite, and halloysite. Except for halloysite, the other minerals are nonexpandable in water. Halloysite contains interlayer water, as will be discussed in the following pages. Upon heating, it is irreversibly dehydrated and the mineral is called metahalloysite. Of the mineral species listed in the foregoing, kaolinite is perhaps the most widely distributed in soils. It is an important fraction of the clay of Ultisols and Oxisols and is also detected as accessory minerals in Alfisols and Vertisols in the tropics.

Halloysite (1:1 Layer Clays)

Hallosite has a general composition $Al_2O_3 \cdot 2SiO_2 \cdot 4H_2O$, and is similar in structure to kaolinite. The differences lie in the disordered stacking of layers and in the presence of two or more interlayers of water, as noted earlier for this mineral.

The water molecules are linked together in a hexagonal pattern.

In turn, they are bonded to the crystal layers by H bonding. Because of the presence of interlayer water, halloysite exhibits a basal (001) spacing of 10.1 Å, which upon heating can be reduced to 7.2 Å (Figure 5.4). The dehydrated species is called metahalloysite, Halloysite is reported to convert rapidly into metahalloysite at 50°C, but the basal d spacing will collapse only after heating at 400°C. Although heating reduces the d spacing, it does not affect the random stacking of layers.

Halloysite, in general, is tubular in form, as shown by electron microscopy (Figure 5.5). This is in contrast to kaolinite, which is hexagonal. However, recently sheetlike halloysite, called *tabular* halloysite, has been detected in some of the soils in Texas. The tubular crystal form is considered to be rolled up sheets. Because of the presence of interlayer water, the normal z-axis bonding of O—OH groups is prevented, causing a distortion of the crystal structure with the consequent curling of layers.

The x-ray diffraction pattern of halloysite, dried at 105°C, is almost similar to kaolinite. However, the basal (001) diffraction peak of halloysite is usually broad or less sharp, owing to the disordered stacking of layers. Partially dehydrated halloysite may exhibit x-

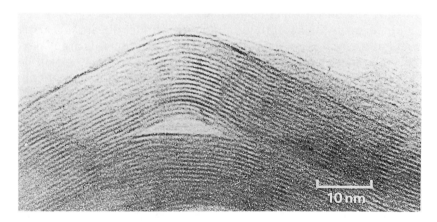

Figure 5.4 Transmission electron microscopy of halloysite showing lines corresponding to basal unit layers, repeating at distances of approximately 7.2 Å (0.72 nm). (From Sudo and Yotsumoto, 1977; Courtesy of Clays and Clay Minerals.)

Figure 5.5 Transmission electron micrographs of kaolinite (left) and halloysite (right) minerals. (From Egawa and Watanabe, 1964.)

139

ray diffraction patterns between the two end members stated earlier (between 10.1 and 7.2 Å).

Formation and stability of halloysite in soils appear to be influenced by soil moisture. A moist condition is considered to be required for the development of this mineral. There are indications that halloysite can be considered a precursor of kaolinite, since formation of the mineral follows the weathering sequence: igneous rock → montmorillonite → halloysite → methahalloysite → kaolinite.

Montmorillonite (Expanding 2:1 Layer Clays)

Minerals in this group are sometimes called *smectite* and have a variable composition. However, the formula is often expressed as $Al_2O_3 \cdot 4SiO_2 \cdot H_2O + xH_2O$. The name montmorillonite is reserved for the hydrated aluminosilicate species with little substitution. Many clay deposits in the United States contain large amounts of montmorillonite. This type of clay is frequently called *bentonite*, and commercial grade montmorillonite is also often referred to as bentonite.

A wide range of minerals exist within the montmorillonite group, and the principal end members in the dioctahedral subgroup are beidellite and nontronite. Montmorillonite has Mg and ferric ions in octahedral positions, whereas beidellite ideally contains no Mg or Fe in the octahedral sheet. Beidellite is characterized by a high Al content. The silicate layer charge is derived entirely by substitution of Al^{3+} for Si^{4+}. Nontronite is like beidellite, but with all the Al^{3+} replaced by Fe^{3+}. In the trioctahedral subgroup, only two end members are recognized: hectorite and saponite.

Two types of structure have been proposed for montmorillonite, the structure according to (1) Hofmann and Endell and (2) that of Edelman and Favajee. Both hypotheses show similarity in the fact that the unit cell structure is considered symmetric, as opposed to that of kaolinite. One aluminum octahedral sheet is sandwiched between two silica tetrahedra sheets. The crystal layers are reported to be stacked together in random fashion, and some of the minerals may even be fibrous, such as hectorite. The bonds holding the layers together are relatively weak, developing intermicellar spaces that will expand with increasing moisture content (see figure). However, the difference between the structure of Hofmann and Endell and that of Edelman and Favajee is in the arrangement of the silica tetrahedra network, as shown in Figure 5.6. Edelman and Favajee are of the opinion that an alternative arrangement of silica tetrahedra exists with a Si—O—Si bond angle = 180°, with basal planes composed of OH groups bonded by the silica in the tetrahedrons.

The negative charge of montmorillonite arises mainly from isomorphous substitution. Only a small variable charge is present, since all the disposable hydroxyl groups are located in subsurface planes covered by a network of oxygen atoms. Van Olphen (1977) mentioned a charge equivalent to a cation exchange capacity of 70 mEq/100 g for a typical montmorillonite. The specific surface area is approximately 700–800 m^2/g, and because of this large specific surface area, which is exposed on dispersion in water, montmorillonite exhibits strong plasticity and stickiness when wet.

The minerals are generally very fine grained, whereas the component layers are not bonded strongly, as noted earlier. In contact with water, the mineral exhibits interlayer swelling, causing the volume of the clay to double. Indications are available that the basal spacing of montmorillonite increases uniformly with adsorption of water. Several authors noted that the increase in basal spacing can occur stepwise, suggesting formation of hydration shells around interlayer cations. The high swell–shrink potential is why the mineral can admit and fix metal ions and organic compounds. The adsorption of organic compounds leads to formation of organomineral complexes. Organic ions are believed to be able to replace inorganic cations in the interlayer position. Monolayers, sometimes double layers, of organic molecules are adsorbed, depending on the size of the cations and the charge deficit of the layers.

Adsorption of organic compounds, such as glycerol and ethylene glycol, is diagnostic in identification of montmorillonite by x-ray diffraction analysis. Ovendried (105°C) montmorillonite is usually characterized by a basal (001) spacing diffraction peak at 10 Å. In air–dry conditions, the mineral has some interlayer water, and the characteristic spacing is approximately 12.4–14 Å (Figure 5.7). After intercalation with ethylene glycol or glycerol, the basal (001) d spacing expands to 17.0 Å. Reports are present in the literature that suggest that the spacings can be expanded indefinitely. However, here the mineral exists only as platelets or as unit cells, whereas the intermicellar spaces cease to exist.

Among the several species of clays in soils, the mineral mont-

142

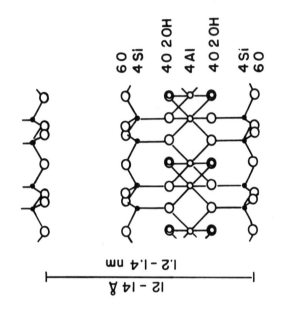

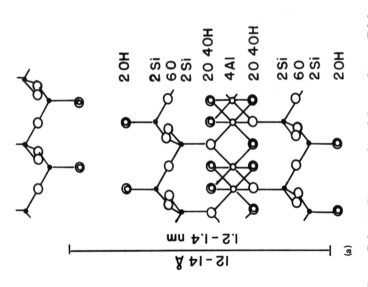

Figure 5.6 (a) Structural model according to Edelman and Favajee, and (b) structural model according to Hofmann and Endell. [From M. W. Wirjodihardjo, and K. H. Tan (1964). *Ilmu Tanah*. Djilid II. Publishers Pradnjaparamita II, Jakarta, Indonesia.]

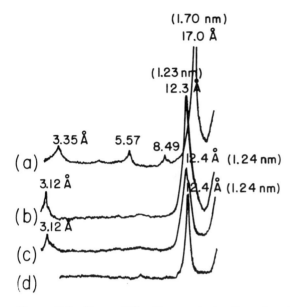

Figure 5.7 X-ray diffraction analysis of montmorillonite: (a) after solvation, (b) through (d) air dry. (From Tan and McCreery, 1975.)

morillonite is perhaps the most widely distributed member. Most soil montmorillonites are dioctahedral. They are characteristic constituents of clays of Vertisols, Mollisols, and Alfisols and are also found in some Entisols. The high plasticity and swell–shrink potential of the mineral make these soils plastic when wet and hard when dry, whereas wide cracks will form as soils dry out. The dry soil is difficult to till.

Illites (Nonexpanding 2:1 Layer Clays)

The illite minerals are micaceous types of clay. However, in contrast with the true micas, which are of primary formation, illites are of secondary origin. They are also known under the names hydrous mica or soil mica. The term *illite* is suggested for the fine-grained minerals in this group, whereas the coarser particles are frequently called *hydrous mica*.

Several authors object to classifying illite as clay. They indicate that, strictly speaking, illite is a clay-sized mica and, therefore, cannot be regarded as a clay mineral (Fanning and Keramidas, 1977; Theng, 1974). However, illitic clay mineralogy is recognized by the U.S. soil taxonomy classification (USDA, 1975). Van Olphen (1977) considers mica, especially muscovite, the prototype of illite. Its close relation with mica was the reason for naming this mineral hydrous mica or soil mica.

The mineral has a chemical composition almost similar to muscovite, but contains more SiO_2 and less K. Alteration, associated with weathering processes of muscovite and subsequent formation into clays, are the reasons for the observed differences. Several authors are of the opinion that a continuous series of illite species exists between muscovite and montmorillonite as the end members. Mixed layering of illite–montmorillonite often occurs.

$$H_2KAl_3Si_3O_{12} \longleftrightarrow Al_2O_3 \cdot 4SiO_2 \cdot H_2O + xH_2O$$

| Muscovite | illite series | Montmorillonite |

Since illite contains interlayer potassium, the unit layers are bonded more strongly than montmorillonite. Therefore, the intermicellar spaces of illite do not expand upon addition of water. Because of the electrostatic bonds exerted by K^+ ions linking the unit layers together, the basal (001) spacing is 10 Å. The cation exchange capacity is about 30 mEq/100 g, and plasticity, swelling, and shrinking are less intense in illite than in montmorillonite. Fine-grain illites have been found concentrated in the coarse clay fraction (2– 0.2 μm) of soils. As indicated earlier, the physical properties are closer to kaolinite than to montmorillonite. However, the ease of parallel alignment of the particles and their presence in the coarse clay fraction are considered to have a detrimental effect on soil stability.

Identification of illite can easily be made by x-ray diffraction analysis. These minerals are characterized by a basal (001) spacing of 10.0 Å. The latter peak does not shift or collapse after heating the mineral at 500°C or after solvation with glycerol, glycol, or ethylene glycol. In many instances, elemental analysis of K concentration has also been used for the detection of illites in soil clays. Although the theoretical concentration is approximately 9–10% (K), a potassium content of 5–8% is frequently found for illitic clays, with a value of 7% (K) as the diagnostic percentage for illite.

Illite is an important constituent of clays in Mollisols, Alfisols,

Spodosols, Aridisols, Inceptisols, and Entisols. In soils affected by high precipitation, the mineral tends to be altered into montmorillonite, whereas under the influence of warmer climates or higher temperatures, the structure of illite is reported to become more disordered, and kaolinite is formed.

Vermiculites

The vermiculite group of minerals also forms micalike flakes as illites. It is also a mica alteration product. However, the mineral called *hydrobiotite*, formed from weathering of biotite, is not vermiculite, but belongs to the illite group.

Vermiculite can be divided into two categories: true vermiculite and clay vermiculite. True vermiculite is not considered a clay mineral, but a rock-forming mineral (Douglas, 1977; Walker, 1975). The name is derived from *vermiculare* or *vermicularis* (Latin, wormlike, or to breed worm), since upon heating the mineral becomes elongated, twisted, and curved. After heating, it usually expands to 20–30 times its original size. Commercial vermiculite is often interlayered biotite and vermiculite. The clay-sized vermiculite found in soils is considered "clay vermiculite" or "soil vermiculite." Its existence in the clay fraction of soils was first demonstrated in 1947 in the soils of Scotland, but Walker (1975) indicated that it has not been isolated yet as monomineral particles. The detection in soils is based on its x-ray diffraction peak at 14 Å; this is why it is sometimes called 14-Å mineral.

Clay vermiculite is a magnesium aluminum silicate, with Mg occupying the octahedral positions between two silica tetrahedra sheets. Some iron may also be present. The chemical formula can be generalized as

$$22MgO \cdot 5Al_2O_3 \cdot Fe_2O_3 \cdot 22SiO_2 \cdot 40H_2O$$

or

$$Mg_3Si_4O_{10}(OH)_2 \cdot xH_2O$$

The structure (Figure 5.8) shows close similarities with that of chlorite, with the difference that, instead of brucite, water molecules of about 5-Å layers are occupying the intermicellar spaces. In many cases, interlayering with hydroxy-Al also occurs. In the tetrahedral

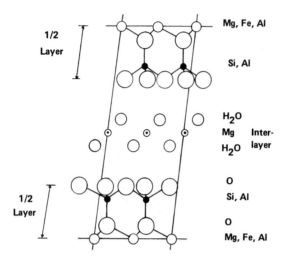

Figure 5.8   The structure of vermiculite, showing the brucite interlayer.

layer considerable substitution of Al for Si takes place. This is the reason for the high negative charge present in vermiculite.

Vermiculite is one of the clay minerals with the largest CEC among the inorganic colloids. The cation exchange capacity is approximately 150 mEq/100 g and exceeds that of montmorillonite. According to Douglas (1977), the CEC of diocathedral vermiculite is 1.05 times the CEC of trioctahedral vermiculite, and values of CEC between 144 and 207 mEq/100 g have been reported. The presence of hydroxy–Al interlayers usually reduces the CEC of the mineral. Vermiculite with hydroxy–Al in intermicellar spaces is presently called *hydroxyaluminum interlayer vermiculite*, or *HIV*. In the older literature, this type of vermiculite is known as *interstratified vermiculite*, because of the presence of α-$Al(OH)_3$, hydroxy–Al, or gibbsitic–Al, in interlayer position. Soils of the southern region in the United States, often characterized by low CECs, exhibit higher CEC values by small admixtures of vermiculite, or HIVs, in their clay fraction.

Most soil vermiculites are probably dioctahedral. The mineral is reported to have wedge zones with high selectivity for fixation of K^+, NH_4^+, and other cations. The high potassium and ammonium fixation values in many soils are attributed more to the presence of vermiculite than to montmorillonite or illitic type of clays.

Identification of clay vermiculites is done mostly by x-ray diffraction analysis and differential thermal analysis (DTA). For oriented samples, the basal (001) x-ray diffraction peak is 14 Å. This peak will not shift or collapse upon solvation, but after heating to 700°C, the basal d spacing usually collapses to 11.8 or 9.3 Å. The common occurrence of vermiculite as mixed layers with montmorillonite, chlorite, illite, and biotite, yields many difficulties in the positive identification of vermiculite. Many clay vermiculites may have been identified in the past as montmorillonites. In some instances, treatment of vermiculite with KCl solutions can produce a mineral with a mica structure.

Vermiculite usually occurs as accessory minerals in the clay fractions of Ultisols, Mollisols, and Aridisols. It is formed more in well-drained soils, in contrast with montmorillonite, which requires a gley condition for formation.

Chlorites (2:2 Layer Clays)

Chlorites are hydrated magnesium and aluminum silicates, which are related to mica minerals in appearance. The name comes from the green color of many chlorite specimens. Structurally chlorite is related to talc, or 2:1 layer clays, and shows a close relationship with vermiculite. However, recently, a number of authors prefer to use the term *2:2 layer* for chlorite. Octahedral sheets, composed of $Mg(OH)_2$ are sandwiched between the two silica tetrahedra sheets. The $Mg(OH)_2$ sheet was formerly called the *brucite* sheet. The intermicellar spaces are also occupied by brucite sheets; hence, the term 2:2 layer clays.

The mineral is variable in composition, but the general composition is reported as follows:

$$(Mg, Fe, Al)_6(Si, Al)_4O_{10}(OH)_8$$

Isomorphous substitutions occur in both the tetrahedral and octahedral layers. The silicon may be replaced by Al, whereas Fe or Al may replace Mg in octahedral positions. The degree of substitution is expressed by Foster (1962) as Fe^{2+}/R^{2+} ratios and, on this basis, three general groups of chlorites are recognized (Figure 5.9): (1) Fe chlorites, containing relatively high amounts of iron, (2) intermediates, and (3) Mg chlorites, which contain smaller amounts of iron.

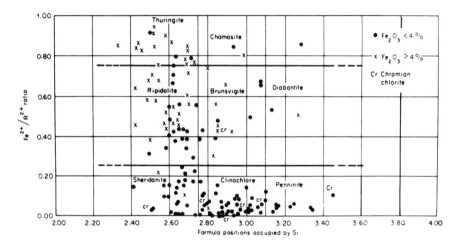

Figure 5.9 Classification of chlorites on the basis of degree of isomorphous substitution as expressed in Fe^{2+}/R^{2+} ratios. (From Foster, 1962.)

Other types of chloritic minerals have also been recognized (e.g., swelling chlorites or corrensites). The latter swells when wet. They are supposed to be more mixed-layer minerals composed of chlorite–montmorillonite and vermiculite than normal chlorite.

The replacement of Mg by Al occurring in the brucite sheets accounts for the development of a positive charge. This positive charge practically neutralizes the negative charge of the "mica" layer. Therefore, chlorite has only a very small charge and, consequently, a small CEC. The hydroxy interlayers are sites for anion retention. Phosphorus is reported to be fixed by interlayer hydroxides of chlorites. On the other hand, the presence of these interlayers reduces fixation of K or NH_4^+ ions.

Depending on the species, the characteristic d spacing (001) of chlorite is 14.0 Å, as determined by x-ray diffraction analysis using oriented specimens. This peak does not shift or collapse by treatments of the sample with glycerol or ethylene glycol or by heating to 500°C. Swelling chlorite may have a basal spacing of 28 Å, which increases to 32 Å by solvation.

Abundance and frequency of occurrence of chlorite are considered low (Barnhisel, 1977). Chlorite is usually detected as accessory minerals in clays of Alfisols, Mollisols, and Aridisols. Most of the chlorite minerals are trioctahedral, but recently, dioctahedral chlorite has been detected in Virginia from soils derived from muscovite-schist and in soils from British Columbia.

Mixed-Layer Clays

Soil clays exist in nature as a mixture of several species. Some of the different types of clay minerals may be stacked together as a packet. This is called *interstratification*, and such clays are referred to as *interstratified* clays or mixed-layer clays. Interstratified clays cannot be separated by physical means, as can an ordinary mixture of clays.

Interstratification can occur in a (1) regular fashion or (2) random fashion. It may also be the result of a segregation process within a crystal in zones within another mineral. Another process of formation of mixed-layer clays is precipitation, formation, or crystal growth in interlayers. For example, gibbsite sheets may develop from precipitation and crystallization in intermicellar spaces, owing to replacement of exchangeable cations and change in chemical environment.

The identification of regular mixed-layer clays follows the same principles as used with monomineral clays. X-ray analysis also produced a regular sequence of x-ray diffractions with regular mixed clays. They are usually identified by an integral sequence of the basal (001) diffraction peaks, which corresponds to the sum of the thickness of the component layers. For example, two layers of 14 Å vermiculitic clay will yield a basal d spacing of $2 \times 14 = 28$ Å. Corrensite is considered a regularly interstratified chlorite–montmorillonite clay. Here, the basal (001) diffraction peak is either $14.0 + 17.0 = 31.0$ Å (for an expanded montmorillonite component), or $14.0 + 12.4 = 26.4$ Å (for an air–dry montmorillonite component), assuming that the structure has not been changed.

The random mixed-layer clays are more difficult to identify, and do not exhibit integral series of basal (001) diffraction patterns. The sequence of diffraction is relatively short, compared with regular mixtures. Pretreatments, such as solvation, potassium saturation, and heating, may then be required to solve the problem.

On the other hand, physically mixed clays can be readily identified, since the diffraction of the basal (001) spacings of the major planes will all appear in the x-ray analysis. Each of the diffraction peaks can be identified in the usual manner.

Interstratified clays have been detected in a large variety of soils in temperate, cold, and tropical regions. MacEwan and Riuz-Amil (1975) are of the opinion that mixed layering is less common in tropical conditions. In the soils of the humid temperate regions, interstratification occurs often in the sequence montmorillonite–chlorite–mica, or mica–illite. However, montmorillonite–kaolinite, or vermiculite–kaolinite has also been observed, especially in the subtropical regions of the United States. In the humid tropics, interstratification has been noted in the sequence montmorillonite–halloysite–kaolinite. In the United States, chlorite and vermiculite are the most common interstratified clays in Alfisols and some Ultisols.

Silica Minerals

Silica minerals are minerals composed entirely of silica. They occur extensively in nature and are frequently an important constituent of the clay fractions of soils. However, the coarse silica particles are found mostly in the silt and sand fractions.

Structurally these minerals do not belong to the phyllosilicates characterized by sheet structures, but they are distinguished as minerals with framework structures or tectosilicates. The four oxygen atoms of the silica tetrahedron are linked directly to neighboring silicon atoms, yielding a fourfold coordination that is electrically balanced. The formula of these minerals is generalized as $n(SiO_2)$. Three types of minerals are distinguished in the category of silica minerals: quartz, tridymite, and crystobalite. Depending on the temperature, each of them can exist in an α and β. The α-form is the low-temperature variety, whereas the β-modification is the high-temperature form. The transformation, called *conversion*, is normally instantaneous and reversible, and accompanied by structural changes (Figure 5.10).

α-Quartz (trigonal) α-tridymite (hexagonal)
$\updownarrow$ 573°C $\updownarrow$ 117°C
β-Quartz (hexagonal) $\rightleftharpoons$ β-tridymite (hexagonal)
                          870°C

                              α-crystobalite (tetragonal)
                              $\updownarrow$ 220–280°C
                    $\rightleftharpoons$ β-crystobalite (cubic)
              1470°C

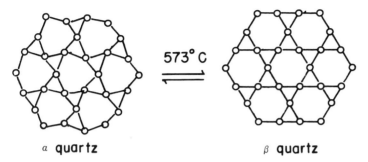

α quartz   β quartz

Figure 5.10 Structural changes associated with conversion of α- to β-quartz.

The silica minerals are generally considered inert, or chemically inactive, material. They have only a slight effect on physicochemical properties of soils, and because of their low chemical activity, they are perhaps of importance only as a diluent to the more reactive clay and humic material. The surface area is very small, and amounts to only 2–3.0 m^2/g, depending on the shape of the particles. Soils with clay fractions dominated by silica minerals, are usually nonplastic and have a small shrink–swell capacity, as well as a small water-holding capacity. The surface charge is also small, if not negible, whereas the correspondingly small cation and anion exchange are more attributable to the Si-O broken bonds and Si—OH groups on particle edges.

Silica minerals are insoluble at low pH. Their solubility does not increase if pH is increased (e.g., from pH 3 to pH 9; Krauskopf, 1956). Only above soil pH 9.0, will silica dissolve according to the following reaction:

$$Si(OH)_4 + OH^- \rightleftharpoons Si(OH)_3O^- + H_2O$$

In general, solubility of silica minerals is related to the packing density of the silica tetrahedra. Of the three types of minerals stated, solubility increases in the following order:

Quartz < crystobalite < opal < amorphous silica

The sequence suggests that quartz is the least soluble, whereas amorphous silica is the most soluble form. Opal is mostly of plant origin and is also called *biogenic silica*.

Identification of silica minerals is usually done by x-ray diffraction and DTA. With x-ray analysis quartz yields a d (100) spacing of 4.26 Å and a d (101) spacing of 3.34 Å. Frequently the d (100) diffraction peak at 4.26 Å is very weak in intensity, leaving the usually strong 3.34 Å x-ray diffraction peak for use in diagnosis of quartz. With DTA, quartz exhibits a small, but sharp, endothermic peak at 573°C. Since the temperature at which this endothermic peak occurs is sharp at 573°C, quartz is often used as a stable reference material for calibration of the DTA instrument. The identification of crystobalite is more difficult than quartz, since crystobalite gives a series of d spacings by x-ray analysis (e.g., 4.04, 3.14, and 2.84 Å), which overlap the d spacings of orthoclase minerals. Generally, a strong 4.04-Å peak accompanied by a relatively weak 3.14-Å peak suggests the presence of crystobalite.

The silica materials occur in a wide variety of soils. Their contents appear to be related to parent material and to the degree of weathering. Quartz is usually an important mineral in Spodosols because of the formation of these soils from highly siliceous material. Inceptisols may be rich in quartz, but this may also be a reflection of the parent material. In moderately weathered Alfisols, Ultisols and Mollisols, quartz may accumulate in the eluvial horizons. On the other hand, quartz may be absent in the clay fraction of the highly weathered Oxisols. Crystobalite is often volcanic in origin, and its presence is considered of importance in many volcanic ash soils.

Iron and Aluminum Hydrous Oxide Clays

This group of clays is currently becoming increasingly important. They do not belong to the phyllosilicates, but are oxides of iron and aluminum, containing associated water.

Two major forms of crystalline monohydrates of ferric oxide are known—goethite and lepidocrocite—and two crystalline anhydrous ferric oxides have also been found in soils: hematite and maghemite. The composition of these and other hydrous oxide minerals are listed in the following table.

Goethite	α-FeOOH	Diaspore	α-AlOOH
Lepidocrocite	γ-FeOOH	Boehmite	γ-AlOOH
Hematite	α-Fe$_2$O$_3$	Gibbsite	Al(OH)$_3$
Maghemite	γ-Fe$_2$O$_3$		
Ferrihydrite	Fe$_5$HO$_8 \cdot$4H$_2$O or Fe$_5$(O$_4$H$_3$)$_3$		

Limonite has also been frequently mentioned in the literature as an important rusty iron oxide mineral. However, it is currently no longer considered a soil mineral. A newly discovered hydrous iron oxide mineral, ferrihydrite, is now gaining in importance. According to Schwertmann and Taylor (1977), in the past it has been erroneously called "amorphous ferric hydroxide." It was found as a major component of iron ochre sediments in drainage ditches and is suspected to also occur widely in soils. Perhaps the B_s horizons of Spodosols contain clay fractions with ferrihydrite.

Structurally goethite is formed by close-packed oxygen atoms in a hexagonal pattern (Figure 5.11). On the other hand, lepidocrocite has a more complicated structural pattern. Isomorphic substitution of Al or Mn for some of the Fe frequently occurs.

The most common aluminum hydrous oxide in soils is gibbsite. Gibbsite is sometimes also called *hydrargillite*. Less common aluminum hydrous oxide species are bayerite, boehmite, and diaspore.

The composition of gibbsite is usually formulated as $Al(OH)_3$, with a structure made up of layers composed of two close-packed hydroxyl sheets with aluminum located in a sixfold coordination (Figure 5.12). The hydroxyl groups are arranged in a slightly polar position in the structure. The Al^{3+} ions occupy two-thirds of the possible vacant octahedral interstices. The hydroxyl groups of one layer are almost directly opposite the hydroxyl groups of the adjacent layer. The layers are held together by hydrogen bonds between opposite OH groups.

The iron and aluminum oxide minerals are amphoteric; in acid condition they may have a weak electronegative charge, and in alkaline soil condition they may develop an electropositive charge. At certain pH values, the minerals can also be neutral (no charge). The pH value at which the mineral has no charge is called the *zero point charge* (ZPC). This will be discussed more in detail in a later section of this book.

The adsorption capacity of iron minerals ranges from 30 to 300 μmol/g and, according to Schwertmann and Taylor (1977), this compares favorably with cation exchange capacities of silicate minerals. The latter authors distinguish between "nonspecific" and "specific" adsorption or "chemisorption" of ions by iron minerals. *Nonspecific adsorption* is defined as an electrostatic adsorption, whereas *specific adsorption* is related to a covalent-type ion bonding. Specific adsorption occurs with phosphate ions and heavy metal cations, such as Cu, Zn, Mn, and Pb, leading to a reaction called *retention* or

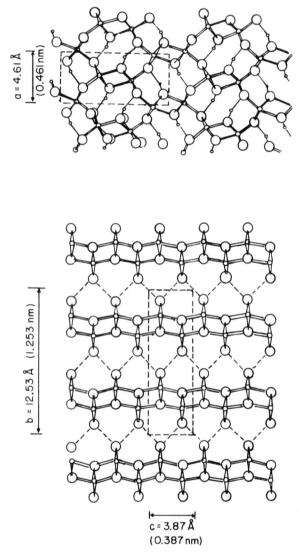

Figure 5.11 Structure of goethite along the *c* axis. (From Greenland and Hayes, 1978.)

a = 4.61 Å
(0.46l nm)

b = l2.53 Å   (l.253 nm)

c = 3.87 Å
(0.387 nm)

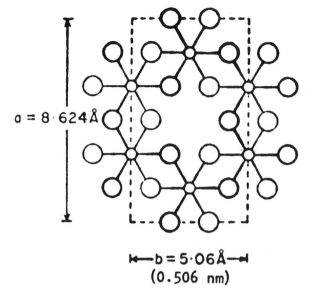

$a = 8 \cdot 624 \text{Å}$

$b = 5 \cdot 06 \text{Å}$
(0.506 nm)

Figure 5.12 Schematic structure of gibbsite. (From Grimshaw, 1971.)

fixation. It has been reported that adsorption of HPO_4^{-2} increases the negative charge of the mineral and, consequently, the CEC will also increase.

The iron oxide minerals have also been known to influence the physical properties of soils. Indications are available that iron oxides are adsorbed on kaolinite surfaces, inducing a cementation effect, leading to the consequent development of strong aggregation of soil particles and to concretion and crust formation (Baver, 1963).

The identification of the iron and aluminum oxide minerals is done by x-ray diffraction and DTA. X-ray analysis yields strong diffraction peaks at 4.18 Å and at 4.82 Å for goethite and gibbsite, respectively. Differential thermal analysis reveals a strong endothermic peak at approximately 290–350°C for goethite, gibbsite, and the other minerals. The problem is usually solved by differential dissolution with NaOH of the clay mixture. This treatment will dissolve the gibbsite components, and an endothermic peak at the temperature range, as indicated in DTA curves of the residue, is then indicative for the presence of goethite or other iron minerals. Scanning electron microscopy shows gibbsite to be rhombohedral (Figure 5.13).

Goethite, hematite, and gibbsite are probably the most frequently found forms of iron and aluminum oxides in soils. They may occur in considerable amounts in the clay fraction, especially of tropical and subtropical soils. Many authors consider their presence to be an indication of the effect of drastic weathering processes. The red and yellow colors of highly weathered soils are attributed to these minerals. Since these minerals can change charges, from electronegative, to zero, and to electropositive charges, or vice versa, they are frequently called clays with variable charges.

Goethite is the most important iron oxide mineral in many soils and is responsible for the yellow to yellowish-brown colors. Hematite, on the other hand, is of less importance, but it may occur in tropical and subtropical soils. This mineral is the reason for the red colors of many tropical soils. Gibbsite is a major mineral in highly weathered Ultisols and Oxisols of the humid tropical and subtropical

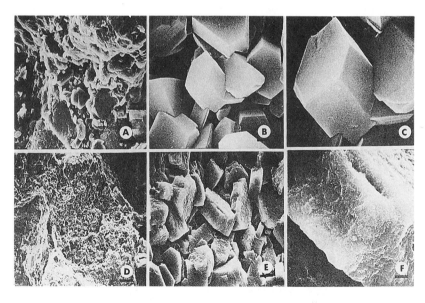

Figure 5.13 Scanning electron microscopy of gibbsite crystals. (A) Magnification ×500, (B) ×5000, (C) ×10,000, (D) ×100, (E) ×1000, and (F) ×5000. (From Eswaran et al., 1977.)

regions. Bauxite deposits in tropical regions contain mostly gibbsite. The U.S. Soil Taxonomy (USDA, 1975; Goenadi and Tan, 1988) uses the gibbsite content in soils as an indication for highly weathered condition. This is expressed in terms of oxidic ratio, which is defined as:

$$\text{Oxidic Ratio} = \frac{\% \text{ Extractable } Fe_2O_3 + \% \text{ Gibbsite}}{\% \text{ Clay}}$$

If the value of the oxidic ratio is ≥ 0.2, the soil is placed in the oxidic class, meaning that the soil has been subjected to drastic weathering (oxidation) processes.

Amorphous Clays, Allophane, and Imogolite

With the recent progress in clay mineralogy, it is currently known that many soils also contain amorphous clays. These clays are non-crystalline and include a wide variety of materials (e.g., silica gel, sesquioxide gels, silicates, and phosphates). They are amorphous to x-ray diffraction analysis. This means that they exhibit a featureless x-ray diffraction pattern. Udo Schwertmann (personal communications) is of the opinion that the term *amorphous* is used erroneously in soil science. It is perhaps the methods of analysis that are inadequate to detect crystallinity in the so-called amorphous clays, since most of these clays occur as very fine, micro, crystal forms. Wada (1977) prefers the use of the term *noncrystalline* over the term *amorphous*.

The most important type of clay in this group is perhaps allophane. It is found especially in volcanic ash soils. The name *allophane* was first introduced by Stromeyer and Hausmann in 1861 for hydrous aluminosilicates occurring in nature. Since then, the name allophane has found general acceptance for a wide variety of clay material amorphous to x-ray diffraction analysis (Ross and Kerr, 1934). Allophane was formerly classified as kaolin clay, since it has a sheet structure similar to kaolinite. Many definitions for allophane are present today (AIPEA, 1992):

1. *Ross and Kerr* (1934) define allophane as an amorphous material that is commonly associated with halloysite. It has no crystal structure and no definite composition, and is a mutual

solution of silica, alumina, water, and minor amounts of bases. Currently allophane associated with halloysite is frequently called *halloysitelike allophane*, with a hypothetical formula of $0.5Al_2O_3 \cdot SiO_2 \cdot 1.4H_2O$. The aluminum in allophanelike halloysite is in octahedral form.

2. *Van Olphen* (1971) defines allophane as a series of naturally occurring minerals that are short-range order hydrous aluminum silicates of various chemical compositions, characterized by Si—O—Al bonds. They exhibit DTA curves with a strong low-temperature endothermic peak and a high-temperature exothermic peak, with no intermediate thermal features.

3. *Farmer et al.* (1985) indicate that allophane is a group name for noncrystalline clay minerals consisting of silica, alumina, and water in chemical combination.

4. *Wada* (1989) reported that allophane is a group name for hydrous aluminosilicates with a composition characterized by a molar Si/Al ratio of $1:2$ to $1:1$. The formula is proposed to be:

$$SiO_2 \cdot Al_2O_3 \cdot 2H_2O \quad \text{or} \quad Al_2O_3 \cdot 2SiO_2 \cdot 2H_2O$$

The mineral consists of hollow irregular spherical particles with a diameter of 3.5–5 nm.

5. *Parfitt* (1990) is of the opinion that allophane is a group name of clay-sized minerals with short-range order. It contains silica, alumina, and water in chemical combination.

Both Van Olphen and Parfitt suggest that allophane is a clay-sized mineral of *short-range order*. Minerals with short-range order have usually long-range disorder, such as glass. X-ray diffraction as well as electron diffraction analysis show no repeat of structural units in any spatial direction. Hence, Parfitt believes that the term *noncrystalline* is more appropriate than the term *short-range order* for allophane.

Another important type of clay in this group is *imogolite*. This clay mineral was reported for the first time in 1962. It was found in weathered volcanic ash or pumice beds, called *imogo* (Yoshinaga and Aomine, 1962). Since then it has been detected in many volcanic ash soils in Japan, South America, and in the islands of the Pacific.

The composition formula of imogolite is assumed to be

$$SiO_2 \cdot Al_2O_3 \cdot 2.5H_2O$$

In many respects, imogolite has chemical characteristics similar to allophane. Several authors believe that allophane is a precursor of imogolite. However, in contrast with allophane, imogolite has a better-defined crystal shape. Electron microscopy shows evidence of the presence of hairlike or spaghettilike crystal forms (Figure 5.14). The term *paracrystalline* has been suggested for the structure of imogolite. The intermediate phase between allophane and imogolite, or imogolitelike allophane is called *protoimogolite*. This is allophane with a structure close to that of imogolite, but lacks the crystal order of imogolite. The suggested chemical formula is: $Al_2O_3 \cdot 0.6-1.0SiO_2 \cdot 2.5-3.0H_2O$. It has a morphology of hollow spherules with an outside diameter of 3.5–5.0 nm. The micrograph of the clay in the volcanic ash soil in Chile (see Figure 5.14, right) is, in fact, protoimogolite, since it is still associated with large amounts of allophane. "True" imogolite often exhibits a dominant hairlike structure, as shown in Figure 5.15.

The presence of allophane gives the soil unique properties. Allophane has a large variable charge. It also behaves amphoterically and is reported to fix considerable amounts of phosphates. The CEC is approximately between 20 and 50 mEq/100 g, whereas the AEC ranges from 5 to 30 mEq/100 g. Imogolite has a larger CEC value than allophane. Wada (1977) estimated the CEC of imogolite to be 135 mEq/100 g clay. As with the iron and aluminum oxide minerals, anion adsorption by allophane is also divided into nonspecific and specific adsorption. *Nonspecific adsorption* also refers to electrostatic adsorption, whereas *specific adsorption* is the adsorption of ions by covalent bonding in the coordination shells of the Al (or Fe) atoms. The amount of ions adsorbed nonspecifically increases with lower pH. Ions adsorbed by the specific process are considered to be fixed or, in other words, they can be replaced only with difficulty by other ions. This fixation process is of special importance in phosphate fixation.

The presence of allophane has also an important effect on several soil physical properties. Soils high in allophanic clays are characterized by low bulk density values, high plasticity, although they

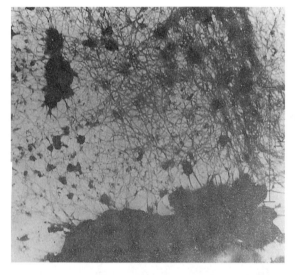

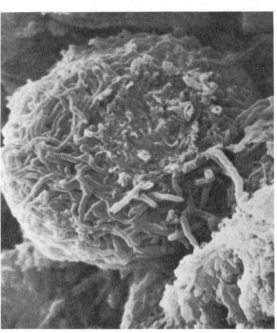

Figure 5.14 Electron microscopy of imogolite. (Left) Kodonbaru soil. (Magnification ×862.) (From Eswaran, 1972.) (Right) Volcanic ash soils in Chile. (From Besoain, 1968.)

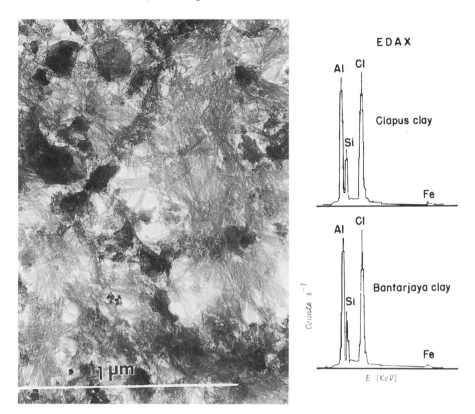

Figure 5.15 Transmission electron micrograph showing the characteristic thin hairlike structure of imogolite and its Al and Si content determined by energy dispersive analysis by x-rays (EDAX). Imogolite was extracted by the method of Yoshinaga and Aomine (1962) from a Ciapus Andosol (Andisol), derived from dacito-andesitic volcanic ash from the Salak volcano, West Java, Indonesia. The Cl detected by EDAX came from NaCl used in the extraction. (From Tan, K. H., personal files, University of Georgia.)

are nonsticky when wet. The water-holding capacity appears to be increased substantially by allophane.

It is assumed that allophane and imogolite will also undergo interaction processes with soil organic compounds, such as humic and fulvic acids. The latter reaction is called *complex formation* or *chelation*. Most soils containing allophane are known to have black A horizons, extremely high in soil organic matter content. Formerly, these soils were called *andosols* or *ando soils* (from the Japanese, *ando*, black).

Identification of these minerals is mostly done by DTA, since x-ray analysis yields featureless diffraction curves. Imogolite exhibits broad x-ray diffraction peaks at 12 Å, at 7.8–8.0 Å, and at 5.5 Å in oriented speciments (Yoshinaga and Aomine, 1962), but the positive identification of imogolite is done by electron microscopy. As shown earlier, imogolite has a hairlike or spaghettilike shape in electron micrographs.

The common method for the detection of allophane is differential thermal analysis. The DTA curve of allophane is generally characterized by a large and sharp endothermic peak between 50° and 200°C (323 and 473 K), attributed to loss of adsorbed water, and a sharp exothermic peak at 900°–1000°C (1173–1273 K), due to formation of γ-alumina or mullite. Imogolite yields by DTA an endothermic peak at 390°–420°C (663–693 K) because of dehydroxylation (Yoshinaga and Aomine, 1962; Wada, 1977), whereas by infrared spectroscopy, imogolite produces an absorption band at 348 cm^{-1} (Farmer et al., 1977; Inoue and Huang, 1990). The importance of acid oxalate-extractable Al and Si have been stressed lately in the characterization of imogolite (Parfitt and Kimble, 1989). It is believed that imogolite exhibited a composition with a characteristic atomic Al/Si ratio of 2.0 (see EDAX data in Figure 5.15). However, in Si-rich soil the Al/Si ratio could be <0.2, whereas in Al-rich soil, minerals with Al/Si ratio >2.0 have also been reported (Farmer et al. 1985; Goenadi and Tan, 1991). This suggests that acid oxalate extraction is a relatively unreliable method for the identification of imogolite. The best method is electron microscopy, which can show the threadlike or fibrous nature of imogolite. The results of other methods are useful as supporting information.

Allophane and imogolite occur mostly in soils of volcanic origin. These soils have been classified under different names in the past (e.g. Andosols, humic allophane soils, Trumao soils, and Kuroboku soils; Tan, 1964). The soils occur extensively in the continents and islands around the Pacific Ocean, and have also been found, to a lesser extent, in the West Indies, Africa, Italy, and Australia. In the United States soils containing allophane are classified first as the Andepts (Inceptisols), but this name has been dropped in 1990 in favor for Andisols.

The presence of imogolite was first reported by Yoshinaga and Aomine (1962) in layers of imogo or pumice beds in Japan. Since then this mineral was detected in many other soils of other countries, as discussed earlier. Although the existence of imogolite was first believed to be associated with volcanic ash soils, the mineral has also been detected in spodic horizons of Spodosols derived from nonvolcanic materials (Tait et al., 1978; Goenadi and Tan, 1991). It has also been synthesized in laboratory conditions (Inoue and Huang, 1990).

5.3 THE IDENTIFICATION OF CLAY MINERALS

In the preceding sections some of the methods for the identification of clays have been mentioned briefly without going into details on the techniques and physicochemical reactions involved. For a better comprehension it is perhaps necessary to discuss briefly three of the major methods often used (e.g., differential thermal analysis, x-ray diffraction analysis, and infrared spectroscopy).

Differential Thermal Analysis

The differential thermal analysis method, commonly referred to as DTA, is a widely used technique and is particularly useful especially in the identification of amorphous material when x-ray diffraction analysis yields only featureless curves (Tan and Hajek, 1977). It found application first in geology and later has been extended to research and analysis in ceramics, glass, polymer, cement, plaster industries, and so forth.

Differential thermal analysis measures the differences in temperature developed between an unknown and a reference sample, as the two are heated side by side at a controlled heating rate from 0° to 1000°C. The reference material, also called *standard material*, is a substance that is thermally inert over the temperature range under investigation. A number of compounds have been used as standard sample [e.g., calcined Al_2O_3 and calcined kaolinite (heated at 1000°C)]. The heating must be controlled at a uniform and steady rate through the analysis. Heating rates may vary from 0.1° to 2000°C/min. For most purposes a heating rate of 20°C/min is used.

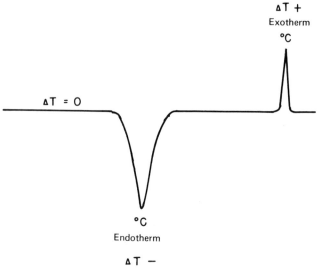

Figure 5.16   Idealized DTA curve.

During the heating process, the unknown sample undergoes a thermal reaction and transformation. The latter is reflected by a difference in temperature between the unknown and reference sample. This difference in temperature is plotted in a graph, usually against the temperature at which the difference occurs (Figure 5.16). If the temperature of the unknown sample becomes lower than that of the reference material, ΔT is negative, an endothermic peak is produced. When the temperature of the sample becomes higher than that of the reference material, ΔT is positive, an exothermic peak develops. The portion of the curve for which $\Delta T = 0$ (no difference in temperature between unknown and reference sample) is considered the baseline. Ideally the baseline is a straight line. Upon analysis by DTA, the mineral may undergo several thermal reactions, culminating in one or a series of endo- and exothermic peaks. The curve with the peaks serves as a fingerprint, and the specific temperatures at which the peaks develop are diagnostic for the identification of the mineral. In addition, the peak height or peak area of the main endothermic reaction can be used for quantitative determination.

Generally DTA can be performed with liquid or solid samples. With soil samples, whole soil, sand, silt, or clay fractions can be used. When whole soils are analyzed, the <2-mm fraction should be treated first with 30% H_2O_2 to remove organic matter, which may interfere by giving strong exothermic reactions. In general analysis, whole soils give only peaks of low intensity. These same peaks are very large and intense if the clay fractions are analyzed (Figure 5.17). However, the quartz inversion peak at 573°C (864 K) is often absent in DTA curves of clays. The sand can be analyzed using the total sand fraction (2.0–0.05 mm) or one of the following sand fractions:

Very coarse sand	2.00–1.00 mm
Coarse sand	1.00–0.50 mm
Medium sand	0.50–0.25 mm
Fine sand	0.25–0.10 mm
Very fine sand	0.10–0.05 mm

DTA of sand shows mostly a strong endothermic peak of quartz at 573°C (846 K). Therefore, it is of importance only in investigation of primary minerals or iron–manganese concretions.

of the main endothermic reaction can be used for quantitative determination.

The clay fraction of soils (<2 μm) can be used directly, or can be separated first, before analysis, into coarse clay (2.0–0.2 μm) and fine clay (<0.2 μm) fractions. For general purposes, the clay fraction <2 μm gives satisfactory results for qualitative and quantitative interpretations. The amount of clay to be used depends on the instruments used. Instruments equipped with well holders need approximately 10–100 mg, whereas those equipped with Pt cups placed on ring-type thermocouples need only 1–10 mg. In qualitative analysis, it is often unnecessary to weigh the sample for DTA, although comparison of curves should be made with curves obtained from identical amounts of samples. On the other hand, in quantitative analysis, the amount of sample must be weighed accurately, since height or area of the main endothermic peak increases proportionally with sample size.

Qualitative identification of minerals can be achieved by using the DTA curves as fingerprints and comparing or matching them

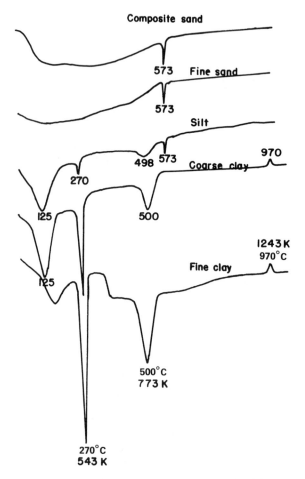

Figure 5.17 Differential thermal analysis (DTA) curves of composite sand (2.0–0.05 mm), fine sand (0.25–0.10 mm), silt (0.050–0.002 mm), coarse clay (0.002–0.0002 mm), and fine clay fractions (<0.2 μm) of a Cecil soil.

with DTA curves of standard minerals or with curves of well-known established minerals. Each mineral exhibits specific thermal reaction features (Figure 5.18). The DTA curves of kaolinite are characterized by a strong endothermic peak at 450°–600°C and by a strong exothermic peak at 900°–1000°C. The endothermic peak is

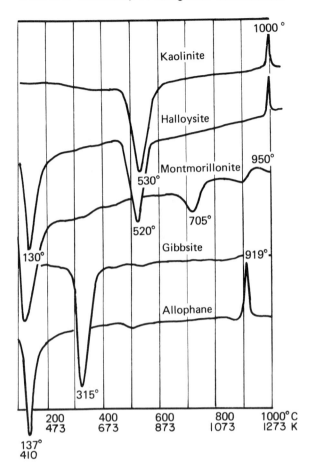

Figure 5.18   Characteristic differential thermal analysis (DTA) curves of selected clay minerals.

caused by dehydroxylation, whereas the exothermic peak is attributed to formation of γ-alumina or mullite or both. The curve of halloysite is almost similar to that of kaolinite, but has in addition a low-temperature (100°–200°C) endothermic peak of medium to strong intensity for loss of adsorbed interlayer water. Montmorillonite exhibits a DTA curve characterized by a low-temperature (100°–200°C) endothermic peak, an endothermic peak between 600° and 750°C, and a small dip between 800° and 900°C followed by a weak exothermic peak between 900° and 1000°C.

Table 5.3 DTA Endo- and Exothermic Peaks of Major Clay Minerals and the Reactions Causing the Peaks

Mineral	Endothermic peak			Exothermic peak		
	°C	K	Main reaction	°C	K	Main reaction
Kaolinite	500–600	773–873	Dehydroxylation	900–1000	1173–1273	γ-alumina formation
Dickite	500–700	773–973	Dehydroxylation	900–1000	1173–1273	γ-alumina formation
Nacrite	500–700	773–973	Dehydroxylation	900–1000	1173–1273	γ-alumina formation
Montmorillonite	100–250	373–523	Lost of adsorbed water	900–1000	1173–1273	Recrystallization
	600–750	873–1023	Dehydroxylation			
Beidellite	100–250	373–523	Lost of adsorbed water	900–1000	1173–1273	Recrystallization
	500–600	773–873	Dehydroxylation			
Nontronite	100–200	373–473	Lost or adsorbed water	900–1000	1173–1273	Recrystallization
	500	773	Dehydroxylation			
Vermiculite	150	423	Lost or adsorbed water	800–900	1073–1173	Recrystallization
	850	1123	Dehydroxylation			
Illite	100–200	373–473	Lost or adsorbed water	920–950	1193–1223	Recrystallization
	600	873	Dehydroxylation			
	900–920	1173–1193	Dehydroxylation			
Chlorite	500–600	773–873	Dehydroxylation	800	1073	
Halloysite	100–200	373–473	Lost or adsorbed water	900–1000	1173–1273	γ-alumina formation
	500–600	773–873	Dehydroxylation			
Gibbsite	250–350	523–623	Dehydroxylation			
Boehmite	570	843	Dehydroxylation			
Diaspore	400–500	673–773	Dehydroxylation			
Goethite	300–400	573–673	Dehydroxylation			
Quartz				573	843	α-to-β inversion
Allophane	50–150	323–432	Lost of adsorbed water	800–900	1073–1173	γ-alumina formation

Source: Mackenzie (1975), Mackenzie and Caillere (1975), and Tan and Hajek (1977).

Gibbsite and goethite are usually characterized by a strong endothermic peak only between 290° and 350°C. Often goethite and the other iron oxide minerals have their endothermic reaction at a higher temperature than gibbsite. Allophane exhibits DTA features with a strong low-temperature (50°–150°C) endothermic peak and a strong exothermic peak at 900°–1000°C. The low-temperature endothermal reaction is attributed to loss of adsorbed water, whereas the main exothermic reaction is caused by γ-alumina formation.

A list of characteristic endo- and exothermic peaks of major clay minerals is provided in Table 5.3.

X-Ray Diffraction Analysis

The x-ray diffraction method is perhaps the most widely used technique in the identification of clays. It is mainly for qualitative analysis, although frequently semiquantitative determination of clays has been carried out. X-ray diffraction analysis is a nondestructive method, meaning that the sample is not affected by the analysis, and can be used for other analyses. However, the method is not applicable to analysis of amorphous or noncrystalline materials.

The basis for the use of x-rays in the investigation of soil clays is the systematic arrangement of atoms or ions in crystal planes. Each mineral species is characterized by a specific atomic arrangement, creating characteristic atomic planes that can diffract (reflect) x-rays. X-rays are electromagnetic radiation of short wavelength. In most crystals, the atomic spacings, or crystal planes, have almost the same dimension as the wavelength of x-rays. Laue was perhaps the first to discover, in 1912, that x-rays can be diffracted by the atoms in a crystal plane, producing characteristic patterns when recorded. This diffraction pattern is used as a fingerprint in the identification of mineral species.

X-rays are produced in a x-ray tube by fast-moving electrons hitting a metal target. The excited atoms in the target emit radiation with a wavelength between 0.01 and 100 Å, the wavelength of Kα and Kβ radiation. Most metals emit wide bands of Kα and Kβ radiation (e.g., Cu target). By using a nickel filter, the Cu Kβ radiation can be blocked or adsorbed and the Cu Kα radiation is then isolated for use in the analysis. If a beam of Cu Kα radiation hits a crystal plane of a mineral (Figure 5.19), the x-rays are scattered by the atoms of the crystal. To have diffraction occurring, reinforcement must take place of the scattered x-rays in a definite direction. Reinforcement of scattered x-rays becomes quantitative only if Bragg's law is obeyed. Bragg's law is defined as follows:

$$n\lambda = 2d \sin \theta$$

where

d = spacings between atomic planes in the crystal

λ = wavelength

θ = glancing angle of diffraction

n = order of diffraction

The true lattice spacing for the (001) basal plane is the d (001) or the d (hkl) spacing.

Bragg's law predicts that all planes in a crystal diffract x-rays when the crystal is inclined at certain angles to the incident beam. The angles θ depend on the wavelength λ and on d. If Figure 5.19 is studied, the incident beam DEF has traveled several integral numbers of wavelength ($n\lambda$) farther than incident beam ABC. Diffraction from a succession of equally spaced atomic planes yields a diffraction maximum. If these diffractions are received by a photographic film, a series of spots or lines (bands) are produced. The position of the lines is related directly to the d spacings.

The value $d(001)/n$, called *lattice spacing*, can be measured from results of x-ray diffraction analysis. If $n = 1$, the $d(001)/1$ represents

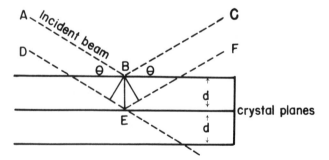

Figure 5.19 Schematic drawing of an x-ray incident beam diffracting from crystal planes, obeying Bragg's law: $n\lambda = 2d \sin \theta$.

the first-order diffraction spacing. If $n = 2$, then $d(001)/2$ represents the second-order diffraction spacing. The series of d/n values obtained, together with the intensity of the x-ray diffraction peaks, are diagnostic for identification of mineral species.

The accepted unit for lattice spacing is the angström unit (10^{-10} m $= 1$ Å), which corresponds to the unit of x-ray wavelength. In some books, the nanometer (1 Å $= 0.1$ nm) is preferred, whereas in older books, the kX unit is used. The kX unit is based on the effective spacing of cleavage planes of calcite ($= 3.0290$ kX). Conversion of kX units into Å units is easily done by multiplying with 1.0020.

The samples for x-ray diffraction analysis can be prepared as a random powder sample or as an oriented sample. In a random powder sample, the crystals lie in a random position to each other. With the aid of glycerol and gum tragacanth the clay sample is made into a paste, and rolled into a rod of 0.3- to 0.5-mm thickness. Powder samples can also be prepared by pushing the paste into specially designed wedge holders. The random powder samples are usually analysed by a powder camera x-ray unit (Figure 5.20). As indicated previously, the diffracted x-rays produce lines or bands on a photographic film. The position of the lines corresponds to the d spacings of the crystal planes of the mineral.

A currently more popular method of mounting samples for x-ray analysis is the preparation of oriented samples on microscopic glass slide, or on porous ceramic plates. A clay suspension is made properly and pipetted onto the slide, so that approximately 15–25 mg of clay is transferred per 10 cm^2. After the sample has been allowed to dry at room temperature, it is ready for analysis with a direct

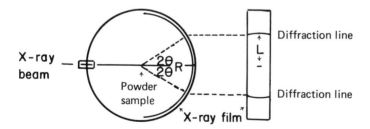

Figure 5.20 Random powder x-ray diffraction camera. R = radius of camera, and L = distance between location of primary beam and diffraction maximum. L is calculated as follows: $L/2\pi R = 2\theta/360°$.

recording x-ray spectrometer, in which the x-ray patterns are printed on charts. The results are normally shown in terms of 2θ values. However, a number of tables are available to convert the 2θ values into d-spacings units. Generally clay minerals exhibit d spacings in the range of 30–3 Å, which corresponds to 2–30° 2θ angles. Highest intensity of diffraction maxima is obtained from d (001) planes. The first-order d (001) diffraction peak, usually together with the second-order diffraction peak, are diagnostic for the identification of the mineral species. The following illustrations serve as a few examples.

Kaolinite (Figures 5.21–5.23) exhibits characteristic first-order diffraction at an angle of 2θ = 12.4°. The latter corresponds after conversion to a d (001) spacing of 7.13 Å. The second-order diffraction is at 2θ = 25.0°, which corresponds to a d spacing of 3.56 Å.

Montmorillonite (air-dry) is characterized by a first-order x-ray diffraction peak of 12.3 Å, which shifts to 17.7 Å after solvation of the sample. The second-order diffraction peak is usually absent.

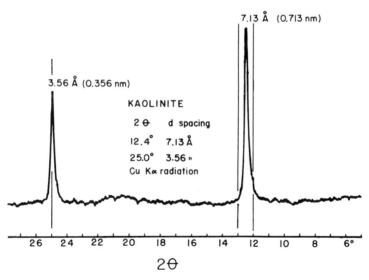

Figure 5.21 Interpretation and identification of kaolinite using its x-ray diffraction pattern. A 12.4 2θ reading corresponds to a d spacing of 7.13 Å (0.713 nm), which is the diagnostic d spacing of kaolinite. See Appendix D for 2θ d spacings values.

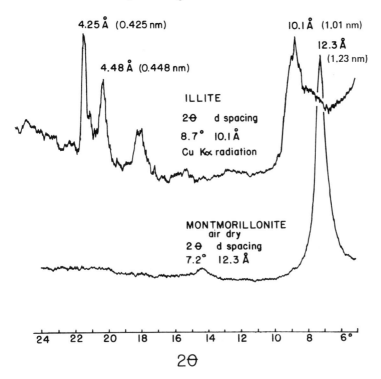

Figure 5.22 Interpretation and identification of illite and montmorillonite using their x-ray diffraction pattern. (Top) A 8.7 2θ reading corresponds to a d spacing of 10.1 Å (1.01 nm), which is diagnostic for illite. (Bottom) A 7.2 2θ reading corresponds to a d spacing of 12.3 Å, which is characteristic for air-dry montmorillonite. See Appendix D for 2θ d spacings values.

Illite exhibits a first-order diffraction peak of 10.1 Å. This peak will not collapse or shift after potassium and magnesium saturation or solvation of the sample.

Gibbsite is identified by the dominant diffraction at approximately 2θ = 18.4°, corresponding to a d spacing of 4.82 Å.

Goethite is easily recognized from the dominant peak at approximately 2θ = 21.6°, corresponding to a d spacing of 4.12 Å.

For additional characteristic d-spacing values of other clay minerals, the following list can be used as a reference (Table 5.4).

As can be noticed from the d spacings, as listed in Table 5.4,

Table 5.4 Characteristic d Spacings of Selected Minerals (Cu Kα Radiation)

d spacings			
Å	nm	Intensity[a]	Clay mineral
17.7-17.0	1.77-1.70	(10)	Montmorillonite, solvated
12.0-15.0	1.20-1.50	(8)-(10)	Montmorillonite, airdry
14.0-15.0	1.40-1.50	(1)	Vermiculite
13.0-14.0	1.30-1.40	(3)-(8)	Chlorites
12.2	1.22		Vermiculites, airdry
11.4-11.7	1.14-1.17	(10)	Hydrobiotite
10.7	1.07		Vermiculites, airdry
10.0-14.0	1.00-1.40		Hydrous mica
10.8	1.08	(10)	Halloysite, hydrated
9.0-10.0	0.90-1.00	(10)	Illites and micas
7.2-7.5	0.72-0.75	(8)	Metahalloysite
7.1-7.2	0.71-0.72	(10)	Kaolinite, dickite, nacrite
6.44	0.644	(6)	Palygorskite
5.90	0.59	(0)-(3)	Montmorillonite, solvated
5.42	0.542	(5)	Palygorskite
5.00	0.500	(9)	Muscovite
4.7-4.8	0.47-0.48	(9)	Chlorite
4.6	0.46	(5)	Vermiculite
4.6	0.46	(10)	Sepiolite
4.4-4.5	0.44-0.45	(9)	Illite, muscovite
4.49	0.449	(8)	Palygorskite
4.45-4.46	0.445-0.446	(4)	Kaolinite
		(6)	Dickite
		(8)	Fireclay
4.43	0.443	(6)	Dickite
4.42	0.442	(10)	Metahalloysite
4.40	0.440	(8)	Nacrite
4.35-4.36	0.435-0.436	(6)	Kaolinite, dickite
4.26	0.426	(4)	Dickite
4.20-4.30	0.420-0.430	(5)	Palygorskite
4.17	0.417	(6)	Kaolinite
4.13	0.413	(6)	Dickite
4.12	0.412	(3)	Kaolinite
3.84	0.384	(4)	Kaolinite
3.82	0.382	(5)	Sepiolite
3.78	0.378	(6)	Dickite
3.69	0.369	(5)	Palygorskite

d spacings			
Å	nm	Intensity[a]	Clay mineral
3.56-3.58	0.356-0.358	(10)	Kaolinite
		(10)	Dickite
		(9)	Nacrite
		(8)	Metahalloysite
	Silicates		
9.20-9.40	0.920-0.940	(9)	Talc
9.10-9.20	0.910-0.920	(6)	Pyrophyllite
7.10-7.20	0.710-0.720	(6)	Antigorite
7.10-7.20	0.710-0.720	(6)	Chrysotile
6.30-6.45	0.630-0.645	(4)-(6)	Feldspars
5.40	0.540	(7)	Mullite
4.60-4.70	0.460-0.470	(6)	Talc
4.57	0.457	(5)	Pyrophyllite
4.00-4.20	0.400-0.420	(8)	Feldspars
3.80-3.90	0.380-0.390	(2)-(7)	Feldspars
3.73-3.75	0.373-0.375	(4)-(8)	Feldspars
3.64-3.67	0.364-0.367	(3)-(8)	Feldspars
3.59-3.60	0.359-0.360	(7)	Antigorite
		(6)	Chrysotile
3.44-3.48	0.344-0.348	(3)-(6)	Feldspars
3.39	0.339	(10)	Mullite
3.36	0.336	(3)	Pyrophyllite
3.10-3.25	0.310-0.325	(7)-(10)	Feldspars
	Oxides and Hydroxydes		
6.25	0.625	(10)	Lepidocrocite
6.23	0.623	(10)	Boehmite
4.96	0.496	(3)	Goethite
4.85	0.485	(3)	Magnetite
4.83	0.483	(10)	Gibbsite
4.72	0.472	(10)	Bayerite
4.62	0.462	(8)	Spinel
4.36	0.436	(8)	Bayerite
4.34	0.434	(6)	Gibbsite
4.29	0.429	(10)	Gypsum
4.21	0.421	(7)	Quartz

Table 5.4 *Continued*

d spacings			
Å	nm	Intensity[a]	Clay mineral
4.15	0.415	(10)	Goethite
4.05	0.405	(10)	Crystoballite
3.98	0.398	(10)	Diaspore
3.84	0.384	(6)	Calcite
3.73	0.373	(7)	Ilmenite
3.72	0.372	(3)	Maghemite
3.67	0.367	(7)	Heamatite
3.36	0.336	(3)	Goethite
3.35	0.335	(10)	Quartz
3.30	0.330	(3)	Gibbsite
3.28	0.328	(9)	Lepidocrocite
3.20	0.320	(6)	Bayerite
3.16	0.316	(10)	Boehmite
3.15	0.315	(4)	Crystoballite
3.06	0.306	(7)	Gypsum
3.30	0.330	(10)	Calcite

[a]Numbers in parentheses refer to intensity on a scale of 1 to 10 (with 1 equaling weak).

several of the minerals have similar, or overlapping, diffraction peaks. In such cases, pretreatments of samples are required to distinguish these minerals. Four major methods frequently used for pretreating the samples prior to analysis are (1) K saturation, (2) Mg saturation, (3) solvation of Mg-saturated samples, and (4) heating at 500°C. The first three methods stated have the purpose of distinguishing between expanding and nonexpanding minerals (Table 5.5). Potassium saturation will normally effect a collapse of intermicellar spacings, and d spacings of 20–17 Å, as exhibited by expanding montmorillonites, may collapse to 10 Å. Reconstitution of the d spacing to 17 Å can be achieved by solvation. None of these treatments will have any effect on the d spacings of nonexpanding minerals (e.g., kaolinite). Heating at 500°C is usually done to distinguish among vermiculite, chlorite, and kaolinite. Vermiculite may have interlayer hydroxy alumina complexes in its structure. These interlayers will be destroyed by heating at 500°C, with the

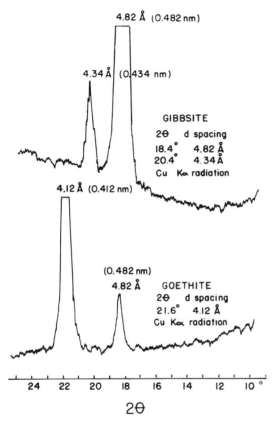

Figure 5.23 Interpretation and identification of gibbsite and goethite using their x-ray diffraction patterns. (Top) A 18.4 2θ reading corresponds to a d spacing of 4.82 Å (0.482 nm), which is the major diagnostic peak for gibbsite. (Bottom) A 21.6 2θ reading corresponds to a d spacing of 4.12 Å (0.412 nm), characteristic for goethite.

consequent collapse of the d spacing of vermiculite from 14 Å to 10 Å. On the other hand, heating at 500°C will show no effect on chlorite, but will produce a 7.2-Å (second-order) diffraction peak. The latter can be confused for kaolinite. However, heating at 500°C will make kaolinite amorphous to x-ray diffraction, and both the 7.13- and 3.56-Å peaks of kaolinite will disappear.

Table 5.5 Effect of Pretreatments on d Spacings of Selected Clay Minerals

d spacings		
Å	nm	Minerals (air-dry)
		K-saturated samples
14	1.4	Vermiculites, chlorites
10–12	1.0–1.2	Montmorillonites, illites
7.2–7.5	0.72–0.75	Hallpysites, metahalloysites
7.15	0.715	Kaolinites, chlorites
		Mg-saturated samples
14	1.40	Vermicullites, chlorites, montmorillonites and illites
10–12	1.00–1.20	Illites, halloysites
7.2–7.5	0.72–0.75	Kaolinites, chlorites
		Solvated Mg-saturated samples
17–18	1.70–1.80	Montmorillonites
14	1.40	Vermiculites, chlorites
10–12	1.00–1.20	Illites, halloysites
7.15	0.715	Kaolinites
		Heated at 500°C (773 K)
14	1.40	Chlorites
10	1.00	Vermiculites
7.0	0.700	Chlorites (kaolinite becomes amorphous)

Infrared Spectroscopy

Recently, infrared spectroscopy has found extensive applications in clay mineralogy studies. Amorphous as well as crystalline clays absorb infrared radiation, and the method can be used when x-ray analysis makes identification difficult. The infrared absorption spectrum of a mineral has a characteristic pattern, which not only permits the identification of the mineral, but also reveals the presence of major functional groups within the structure of the particular compound under investigation.

Infrared absorption is related to molecular or atomic vibrations, and only radiation with a similar frequency as that of the vibration will be absorbed. Atoms and molecules within a compound oscillate or vibrate with frequencies of approximately 10^{13}–10^{14} cps. These frequencies correspond to the frequencies of infrared radiation, and infrared radiation can, therefore, be absorbed by molecular vibrations when the interaction is accompanied by a change in dipole moment. A rapid vibration of atoms yields a rapid change in dipole moment, and absorption of infrared radiation is intense. On the other hand, a weak vibration of atoms produces a slow change in dipole, and, consequently, absorption of infrared radiation is relatively weak. Symmetric molecules will also often not absorb infrared radiation.

Molecular or atomic vibrations cause the interatomic distance to change because the atomic movement, called *oscillation*, subjects the atoms to a periodic displacement relative to one another. The frequency of vibration obeys the law of simple harmonic motion formulated as

$$V = \frac{1}{2\pi c} \sqrt{\frac{k}{m}}$$

where

V = frequency of vibration in cm^{-1}

c = velocity of light in vacuum

m = reduced mass of the vibrating atoms

k = force constant in dyn/cm

Two types of vibrations are distinguished: (1) stretching vibrations, or deformation, in which the atoms are oscillating in the direction of the bond axis without changing bond angles; and (2) bending vibrations, in which the movement of atoms produces a change in bond angles. The restoring force acting on stretching vibrations is usually greater than that required to restore bending vibrations. Therefore, stretching vibrations occur at higher frequencies than bending vibrations. The highest frequencies observed in minerals are those of the stretching vibrations of hydroxyl, OH, groups that occur between 3700 and 2000 cm^{-1}. Bending vibrations occur at lower frequencies, from 1630 to 400 cm^{-1}.

Liquid, gas, and solid samples can be used in infrared spectroscopy. Liquid samples are usually pipetted or injected into infrared cells provided with a NaCl or KBr crystal window. Gas samples are also introduced in cells, similar to the aforementioned cells above. Infrared gas cells are larger than cells for liquid samples and ensure better interaction between infrared radiation and the gas by providing a longer path length. Both NaCl and KBr are infrared inactive and will not interfere in the analysis.

Solid samples should be ground to approximately <2 μm, since coarse particles tend to produce scattering of infrared radiation. Clay fractions (<2 μm) separated by mechanical analysis can be used directly, or can be separated first by centrifugation into coarse (2.0–0.2 μm) and fine (<0.2 μm) clay fractions. If grinding is necessary, it should be carried out with care, since vigorous grinding tends to destroy the mineral structure (becomes amorphous) and tends to increase the hygroscopic nature of the sample.

Several methods have been proposed for mounting solid samples in infrared analysis: (1) mull method, (2) KBr pellet technique, or (3) clay film technique on demountable cells or other support material. The most widely used method is the KBr pellet technique, by which a weighted sample (1–10 mg) is carefully ground with 100 mg KBr, and pressed into a clear transparent pellet. Sodium chloride also appears to be suitable for use in the pellet method. Currently, the use of clay films has attracted considerable attention. The present author noticed that clay films prepared on infrared cells, such as Irtran II (ZnS crystal), or NaCl crystals, give better infrared resolutions than clay samples mounted by the KBr pellet technique (Figures 5.24 and 5.25). However, one disadvantage of the clay film technique is that it takes more time to prepare a clay film than to make a pellet.

In the clay film-mounting method, clay or soil samples are made into a suspension by sonification. They are then pipetted onto Irtan-II window cells, so that 1 mg/cm^2 or 5 mg clay/cm^2 are transferred onto the cells. After drying at room temperature, the cells are scanned from 4000 to 600 cm^{-1} or lower.

Band positions in infrared analysis are indicated in units of frequencies, expressed in terms of centimeters. The *frequency V*, also known as the *wave number*, is defined as the number of waves, or wavelengths, per centimeter (cm^{-1}). It is related to the wavelength

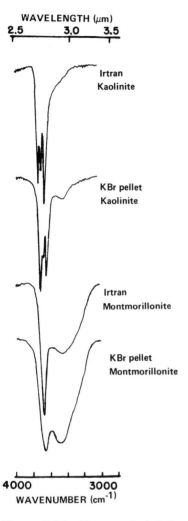

WAVELENGTH (μm)

2.5   3.0   3.5

Irtran
Kaolinite

KBr pellet
Kaolinite

Irtran
Montmorillonite

KBr pellet
Montmorillonite

4000   3000
WAVENUMBER (cm^{-1})

Figure 5.24 Characteristic infrared features in the group frequency region of kaolinite and montmorillonite.

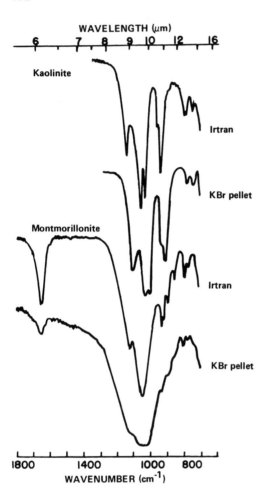

Figure 5.25 Characteristic infrared features in the fingerprint region of kaolinite and montmorillonite.

λ as follows:

$$V = \frac{10^4}{\lambda}$$

in which V is expressed per centimeter (cm^{-1}) and λ in micrometers (μm). The region often analyzed by infrared spectroscopy is in the range of $4000-600\ cm^{-1}$ (equivalent to $2.5-25\ \mu m$) or lower. In many instances, the results of infrared analysis are recorded in the transmittance mode. The latter then yields curves that have an upside down appearance when compared with absorption curves (Tan et al., 1978).

Two groups of frequency regions usually characterize the infrared curves of most clay minerals:

1. Region between 4000 and 3000 cm^{-1} attributed to stretching vibrations of adsorbed water or octahedral OH groups, called the *functional group region.*
2. Region between 1400 and 800 cm^{-1}, attributed to Al—OH or Si—O vibrations, called the *fingerprint region.*

The infrared curves of kaolinite, montmorillonite, and gibbsite, shown in Figures 5.24–5.27, serve as illustrations.

The infrared curve of kaolinite is usually characterized by two strong bands for octahedral OH-stretching vibrations between 3800 and 3600 cm^{-1} when the sample is mounted by the KBr pellet technique. An additional third and very sharp band is present at 3670 cm^{-1} when the samples are mounted as films on Irtran-II cells. The fingerprint region, or the lower frequency region, exhibits sharp bands for kaolinite at 1150, and 1080 cm^{-1} for O—Al—OH vibrations using clay films on Irtran windows. In addition a sharp 1020 cm^{-1} band for Si—O, and sharp bands at 910–920 cm^{-1} for Al—OH vibrations are present. When using KBr pellets, the bands at 1080 and 1020 cm^{-1} appear only as a weakly segregated duplet in most analysis reported for kaolinite.

Montmorillonite also exhibits a better-resolved curve with the clay film technique on Irtran-II. The KBr curve is characterized by one broad band followed by a water band, and one additional dominant broad band at 3640, 3420, and 1050 cm^{-1}, respectively. However, if one uses clay films on Irtran-II, the band at 3640 cm^{-1} for

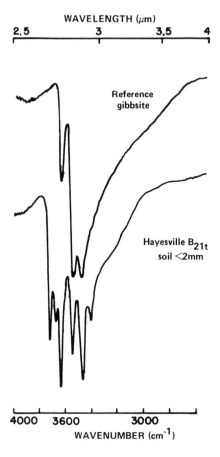

Figure 5.26 Characteristic infrared features in the group frequency region of reference gibbsite and a Hayesville (Ultisols) B_{21t} soil sample containing gibbsite.

OH-stretching vibrations and at 1050 cm^{-1} for Si—O vibrations become very strong and sharp. In addition bands at 1150, 910, 880, and 850 cm^{-1} also increase sharply in intensity in clay film samples.

Standard reference gibbsite (purchased from the Wards Scient. Establ. Co.) is characterized by an absorption band at 3620 cm^{-1}, and by a doublet at 3540 and at 3480 cm^{-1}. In the low-frequency

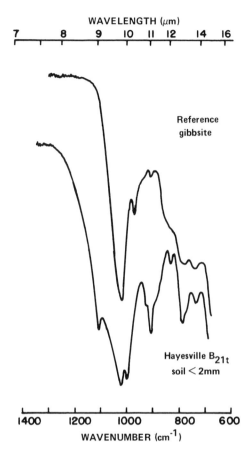

Figure 5.27 Characteristic infrared features in the fingerprint region of reference gibbsite and a Hayesville (Ultisols) B_{21t} soil sample containing gibbsite.

region, gibbsite shows only one dominant peak at 1030 cm^{-1} for O—Al—OH vibration. This peak is the reason for determining the kaolinite band at 1080 cm^{-1} as a separate independent band, rather than calling it a doublet together with the 1020 cm^{-1} band.

Soil gibbsite (see Hayesville B_{21t} soil) exhibits an infrared curve characterized by a triplet between 3600 and 3400 cm^{-1}, instead of the doublet observed for the reference gibbsite. The 3620 cm^{-1} band is overlapping with that of the octahedral OH band of kaolinite in the Hayesville clay fraction. Apparently, the different infrared pattern suggests the presence of different types of gibbsite in soils.

5.4 SURFACE CHEMISTRY OF SOIL CLAYS

Many, if not all, of the chemical reactions of soil clays, are surface phenomena (e.g., cation exchange, adsorption of water). From the preceding section on clay mineral structure, it follows that clay surfaces can be divided into at least three categories:

1. Surfaces formed mainly by Si—O—Si linkages of silica tetrahedrons
2. Surfaces formed by O—Al—OH linkages of alumina octahedrons
3. Surfaces formed by —Si—OH or —Al—OH of amorphous compounds

The first category of surfaces is characterized by surface planes of oxygen atoms, underlaid by silicon atoms of the tetrahedrons. The Si—O—Si bond is called a *siloxane bond* by Sticher and Bach (1966), and because of this, this type of surface can conveniently be called the siloxane surface, and is typical of 2:1 types of clay. The charge of a siloxane surface is mainly attributed to isomorphous substitution of the underlaying silicon atoms of the tetrahedrons.

The second type of clay surfaces is characterized by planes of exposed hydroxyl, OH, groups, underlaid by Al, Fe, or Mg atoms in the center of the octahedrons. Because of the latter, it can perhaps be called the oxyhydroxide surface. Kaolinite and other 1:1 types of clay usually have siloxane surfaces on one basal plane, and oxyhydroxide surfaces on other basal planes. The exposed hydroxyl groups are subject to dissociation and, therefore, play an important role in the development of negative charges.

The third type of surfaces is formed by —Si—OH, called *silanol surfaces*, —Al—OH, called *aluminol surfaces*, and —Fe—OH, called *ferrol surfaces*. These surfaces are typically present in soils containing large amounts of silica gel, amorphous Al- and Fe-oxides, or allophane. The anhydrous Al- or Fe-oxide surfaces become hydrated upon contact with moisture, and aluminol or ferrol surfaces will be developed. The reaction can be illustrated by the following reaction:

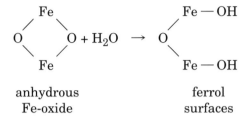

anhydrous   ferrol
Fe-oxide   surfaces

The behavior of these surfaces is expected to be quite different from the other two aforementioned types. Usually, the compounds with silanol, aluminol, and ferrol surfaces have a very large surface area and variable charges, and all the hydroxyl groups are easily accessible.

5.5   SURFACE AREAS

In connection with the surfaces of clay minerals is the problem of surface areas, needed for quantitative interpretation of surface properties in relation to soils and clay behavior. Rate of adsorption and cation exchange are proportional to surface areas. The surface area generally increases with decreased particle size. It can be measured by several methods (e.g., calculation, adsorption analysis, and other procedures). By using the calculation method, the surface area can be measured in terms of (1) total surface area, or (2) specific surface.

Total Surface.   Assume that a cubical container, with side (width) $= L$, is filled with spherical particles. If the dimension of each sphere $= d$ (Figure 5.28), $N =$ number spheres in the container, and $A =$ total surface area of all spheres, then:

$$N = (L/d)^3$$

since: surface area of one sphere $= \pi d^2$, hence:

$$A = N\pi d^2 \quad \text{or} \quad A = (L/d)^3 \pi d^2$$

The total surface area, A, $= \pi L^3/d$ square units

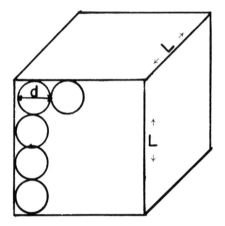

Figure 5.28 Cubical container with sides = L, filled with spheres (balls) with diameter = d.

Specific Surface. The specific surface area, or specific surface, is, by definition, the surface area per unit volume or unit mass of particles.

If the surface area of a sphere = πd^2, and the volume of the same sphere with diameter d equals $1/6\pi d^3$, then by definition:

Specific surface, $S = \pi d^2/(1/6\pi d^3)$ or

$\qquad\qquad S = 6/d$ square units/cubic unit

If we assume that clay particles are spheres with $d = 0.002$ mm, then the specific surface of clay is:

$\qquad S = 6/0.002 = 3000$ mm^2/mm^3

Various types of colloids start to display colloidal properties at different specific surface values. Spangler and Handy (1982) believe that soil constituents begin to exhibit colloidal characteristics when their specific surface reaches values of 6000–10,000 mm^2/mm^3. However, the value of specific surface of 3000 mm^2/mm^3 for the lowest limit at which the colloidal behavior starts to appear, conforms better with the definition of clays (soil constituents with a diameter

Table 5.6 Specific Surface Areas of Selected Clay Minerals

	Total surface area (m^2/g)		
	H_2O method	CPB method	N_2 gas method
Montmorillonite	300	800	784
Mica–smectite (interstratified)	57	152	109
Kaolinite	17	15	32
Allophane	484	0	157

Source: Adapted from Greenland and Quirk (1962, 1964), Dixon (1977), and Wada (1977).

<0.002 mm). The specific surface of the fine clay fraction (<0.0002 mm), which is considered the true colloidal clay, is:

$$S = 6/0.0002 = 30,000 \text{ mm}^2/\text{mm}^3$$

The foregoing calculations and discussion are valid only for uniform particles in the form of perfect balls (spheres). In natural soils, clay is platelike in shape, and its specific surface should be determined with other methods. Its surface area can, for example, be determined by electron microscopy, which is considered the simplest method. Another method is the group of adsorption methods, based on adsorption of compounds in the gas or vapor phase. Depending on the type of reagent used for adsorption by the particles, wide ranges of values are obtained for surface areas. The data in Table 5.6 show some of the variations in surface areas obtained according to adsorption of water, cetyl pyridinium bromide (CPB), or N_2 gas.

5.6 ORIGIN OF NEGATIVE CHARGE IN SOIL CLAYS

As indicated earlier, soil clays ordinarily carry an electronegative charge, which gives rise to cation exchange reactions. This charge is the result of one or more of several different reactions. Two major sources for the origin of negative charges are described in the following sections.

Isomorphous Substitution

Isomorphous substitution is believed to be a major source of negative charges in 2:1 layer clays. Part of the silicon in the tetrahedral layer is subject to replacement by ions of similar size, usually Al^{3+}. In the same manner, part of the Al in the octahedral sheet may be replaced by Mg^{2+}, without disturbing the crystal structure. Such a process of replacement is called *isomorphous substitution*. The resulting negative charge is considered a permanent charge, since it will not change with changing pH.

The ease with which isomorphous substitution takes place depends on the size and valence of the ions involved. It occurs only with ions of comparable size. The difference in dimensions of substituted ions was reported to be no more than 15%, and the valency between those substituted ions should not differ more than one unit (Paton, 1978). In Figure 5.29, the relative dimensions of a few selected ions are given as illustrations. It can be observed that Na^+

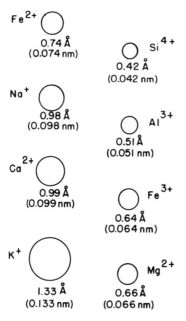

Figure 5.29 Relative dimensions of selected ions commonly found in soils.

and Ca^{2+} are almost of equal size and can replace one another with relative ease, in spite of the larger valence of Ca^{2+}. Potassium is expected to be unable to replace Na^+ or Ca^{2+}, since it is approximately 1.4 times larger than the latter two ions. Magnesium and the iron ions are also of almost equal sizes and may substitute for each other, the sizes being within 15% difference of one another. Aluminum is between Si^{4+} and Mg^{2+} or Fe^{3+} in size and is capable of replacing, with varying degrees of ease, any of these (Paton, 1978).

Dissociation of Exposed Hydroxyl Groups

The appearance of OH groups on crystal edges or on exposed planes, as discussed earlier, can also give rise to negative charges. Especially at high pH, the hydrogen of these hydroxyls dissociates slightly and the surface of the clay is left with the negative charge of the oxygen ions. This type of negative charge is called *variable* or *pH-dependent charge*. The magnitude of the variable charge varies with pH and type of colloid. It is an important type of charge for 1:1 layer, iron, and aluminum oxide clays, and organic colloids; kaolinite also has subbasal hydroxyl groups. Since the latter are surrounded by a network of oxygen atoms from the silicon tetrahedrons, it is expected that the dissociation and consequent contribution to negative charges from subbasal hydroxyl groups may be relatively small.

Not only can protons be dissociated from exposed OH groups, but the latter can also adsorb or gain protons. This process, important only in strongly acidic media, creates positive charges. The reactions for dissociation and association of protons can be illustrated as follows:

Alkaline medium: $—Al—OH + OH^- \rightleftharpoons —Al—O^- + H_2O$

Acid medium: $—Al—OH + H^+ \rightleftharpoons —Al—OH_2^+$

The H^+ and OH^- ions, causing the development of surface charges, are also responsible for the electric surface potential. Therefore, they are called potential-determining ions. The net surface charge will become zero if the negative charge density equals the positive charge density. The pH, at which the latter occurs, is called the *isoelectric point* or the *zero point charge* of the mineral.

5.7 POSITIVE CHARGES AND ZERO POINT OF CHARGE

Soil colloids may also exhibit positive charges as well as negative charges. The positive charges make possible anion exchange reactions and are very important in phosphate retention. These charges are thought to arise from the protonation or addition of H^+ ions to hydroxyl groups. This mechanism depends on pH and the valence of the metal ions. It is usually of importance in Al and Fe oxide clays, but it is of less importance in Si-oxides. For example, gibbsite is positively charged at pH 7.0 or lower.

Gibbsite and other soil colloids may be characterized by a particular pH at which the surface charge is electrically neutral. Previously, this point, or pH value, was called *zero point of charge* pH_0. At pH values above pH_0, the colloid is negatively charged. At pH values below pH_0, the colloid is positively charged.

The zero point of charge (ZPC) can be determined by either the titration method or by analysis of the amounts of adsorbed cations

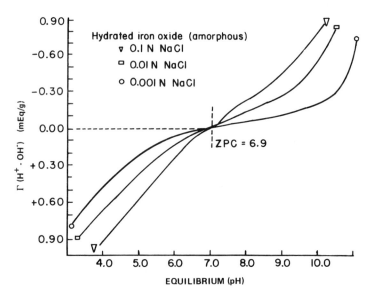

Figure 5.30 Potentiometric titration using NaCl at different strength in the determination of ZPC values of amorphous hydrated iron oxide. (From Gast, 1977.)

and anions as a function of pH and concentrations. If H^+ and OH^- ions are the main potential-determining ions, ZPC is usually found by potentiometric titration and calculations using the formula:

$$\sigma_o = F(\Gamma_{H^+} - \Gamma_{OH^-})$$

where

$$\sigma_o = \text{surface charge density}$$
$$F = \text{Faraday constant}$$
$$\Gamma_{H^+}, \Gamma_{OH^-} = \text{adsorption densities of } H^+ \text{ and } OH^-, \text{ respectively,}$$
$$\text{in mEq/g}$$

The titrations are conducted with an indifferent electrolyte using several concentrations. Figure 5.30 is an example of the presentation of results.

5.8 THE USE OF ΔpH IN THE DETERMINATION OF NEGATIVE OR POSITIVE CHARGES

A relatively simple method to determine whether the net charge of the soil colloids is negative, zero, or positive, is the analysis of soil pH in 1 N KCl and in water. The difference between the two pH values is called ΔpH, and has been used in soil surveys for soil characterization:

$$\Delta pH = pH_{H_2O} - pH_{KCl}$$

The value for ΔpH can be positive, zero, or negative, depending on the net surface charge at the time of sampling and analysis of the soil (USDA, Soil Survey Staff, 1960). A positive value for ΔpH indicates the presence of negatively charged clay colloids. A negative value, on the other hand, means the presence of a positively charged clay colloid. The ZPC is reached when ΔpH equals zero. This, in general, is true if only KCl is present in the system, since KCl is considered to be an indifferent electrolyte. If, however, other anions are present that are subject to specific adsorption they may shift the ZPC value to lower pH regions and, consequently, render the colloid surface more negative in charge (Mekaru and Uehara, 1972).

5.9   SURFACE POTENTIAL

From the foregoing it is clear that if the H^+ and OH^- are the potential determining ions of reversible interfaces, adsorption of protons produces positive charge surfaces, whereas adsorption of OH^- yields negative charges. These reactions can be summarized in the following relationship (Gast, 1977; Van Raij and Peech, 1972):

$$-Al-OH^{0.5-} + H^+ \rightleftharpoons -Al-OH_2^{0.5+}$$

This relationship is of course dependent on pH, whereas at pH_0 (or ZPC) to maintain electroneutrality, the amount of positive charges must equal that of negative charges.

Because of the presence of opposite charges on the colloid surface and in the liquid phase, an electric potential develops at the solid–liquid interphase, called the *surface potential* ψ. The magnitude of the surface potential is given by the Nernst equation:

$$\psi = \frac{RT}{nF} \ln \frac{(H^+)}{(H^+)_{ZPC}}$$

Changing into common log gives

$$\psi = 0.059[\log(H^+) - \log(H^+)_{ZPC}]$$

or

$$\psi = 0.059(ZPC - pH) \text{ volts at } 25°C$$

5.10   ELECTRIC DOUBLE LAYER

Because of the presence of an electronegative charge, clay in suspension can attract cations. These positively charged ions are not distributed uniformly throughout the dispersion medium. They are held on or near the clay surface. Some are free to exchange with other cations. The negative charge of the clay surface is thus screened by an equivalent swarm of counterions that are positive. The negative charge on the clay surface and the swarm of positive counterions are called the *electric double layer*. The first layer of the double layer is formed by the charge on the surface of the clay.

Technically, the charge is a localized point charge; however, we customarily consider this charge to be distributed uniformly over the clay surface.

The second layer of the double layer is in the liquid layer adjacent to the clay surface. The positive counterions in this zone are attracted to the clay surface, but, at the same time, they are free to distribute themselves evenly throughout the solution phase. The two processes will come to an equilibrium, and the resulting distribution zone is similar to the distribution of gas molecules in the earth's atmosphere.

Helmholtz Double-Layer Theory

The Helmholtz double-layer theory is perhaps one of the earliest theories. The negative charge on the colloid is considered to be evenly distributed over the surface (charge density σ). The total countercharge in the second layer is concentrated in a plane parallel

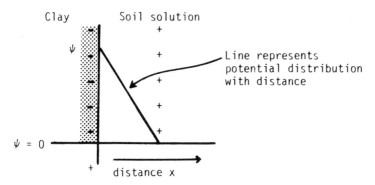

to the surface at distance x. If the medium has a dielectric constant D, then the electrokinetic potential ζ is the same as the total potential ψ:

$$\psi = \frac{4\pi\sigma x}{D}$$

The electrochemical potential is maximum at the colloid surface and drops linearly at locations with increasing distance (x) from the surface within the double layer.

Gouy–Chapman Double-Layer Theory

The negative charge is again considered distributed evenly over the colloid surface. However, the counterions are dispersed in the liquid layer, as are the gas molecules in the earth's atmosphere. This theory is also called the *diffuse double-layer theory* of Gouy and Chapman. The concentration distribution in the liquid zone follows the Boltzmann equation:

$$C_x = C_x^o \exp(-ze\psi/kT)$$

where

C_x = concentration of cations at distance x from surface

C_x^o = concentration of cations in the bulk solution

z = valence

e = electronic charge

ψ = electrical potential

k = Boltzmann constant

T = absolute temperature

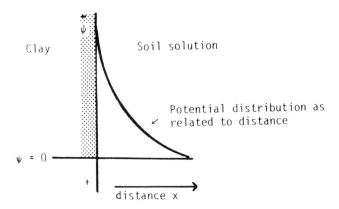

Because of the attraction by the negatively charged surface, cations in the solution phase tend to distribute themselves over the colloid surface so that electroneutrality is maintained, and the tendency for these ions to diffuse away is counteracted by van der Waals attraction. A deficit of anions is usually present in the liquid interface, and the total charge of the surface is considered to be balanced by excess cations. The initial electric potential at the colloidal surface is maximum and decreases exponentially with distance from the surface as follows:

$$\psi_x = \psi_0 \exp(-K_x)$$

where

ψ_x = electric potential at distance x

ψ_0 = surface potential

K = constant associated with concentration, valence of ions, dielectric constant, and temperature

At room temperature,

$$K = 3 \times 10^7 z^{\pm} \sqrt{C}$$

where

z = valence of the ion

C = concentration of the bulk solution in moles per liter

The value $1/K$ is usually used as a measure of the thickness of the double layer (Verwey and Overbeek, 1948). As indicated by the formula for K, the thickness of the double layer is suppressed by both z and C. If C increases with a factor $4x$, K increases with a factor of $\sqrt{4} = 2$. This means that the thickness of the double layer $(1/K)$ is decreased $\frac{1}{2}x$ from the surface. A similar discussion can be given for the valence z.

The Gouy–Chapman diffuse double-layer theory is equally valid for positively charged colloidal surfaces. For a positively charged surface, excess anions will be present in the liquid interface, and a deficit of cations is then expected to occur at the surface.

Limitations to the Gouy–Chapman Diffuse Double-Layer Theory

The diffuse double layer theory was developed independently by Gouy (1910) and Chapman (1913) for the application on flat surfaces, but it may apply equally well to rounded or spherical surfaces (Verwey and Overbeek, 1948). The negative charge was considered to be evenly distributed over the surface. Since the counterions are assumed to be point charges and, therefore, occupy no spaces, they may reach excessively high concentrations at the liquid interface. Modification to the Gouy–Chapman theory was later presented by Stern (1924), who stated that ions of finite sizes cannot approach the colloidal surface more closely than allowed by their effective radii.

Stern Double-Layer Theory

Stern made corrections in the double-layer theory by taking into consideration the ionic dimensions. The influence of ionic dimension is greatest near the colloid. In the Stern theory, the first layer is similar to that of the previous theories. However, the second layer is divided into (1) a sublayer nearest the colloid surface and (2) a diffuse layer:

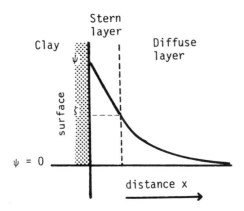

The first sublayer is tightly packed with cations and is called the *Stern layer*. The potential distribution appears to be a combination of the Helmholtz and the Gouy–Chapman diffuse double layer. The decrease in potential is also divided into two parts. In the Stern layer, the potential decreases with distance from the surface according to the Helmholtz theory. From here on (in the diffuse layer), the decrease in potential with distance follows the Gouy–Chapman theory.

Triple-Layer Theory

The double-layer theory was developed for studying adsorption, ion exchange, and other surface reactions on silicate clay surfaces. The negative charge on the clay surface, which is, by definition, the first layer of the double layer, originates from isomorphous substitution within the crystal structure. However, the surface properties of sesquioxides and other types of clays exhibiting oxyhydroxy surfaces, are somewhat different from those of silicate clays. The permanent charge in the oxide clays are complemented or "shielded" on the clay surfaces by charges attributed to adsorption of potential-determining ions, which are primarily H^+ and OH^- ions. Because of the presence of a layer of potential-determining ions, adsorption of the traditional counterions is located in a zone farther outward of the mineral surface. This model, called the *triple-layer model*, is introduced by Yates et al. (1974). Different types of triple-layer models have been presented (Kleijn and Oster, 1983; Bowden et al., 1977). Kleijn and Oster considered the layer of adsorbed potential-determining ions as an integral part of the solid clay surface. This layer, in fact, is located in the liquid phase. It can perhaps be called the *effective* first layer of the double layer:

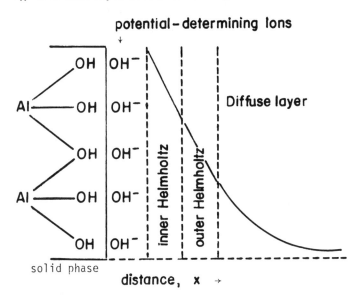

The next layer, also located in the liquid phase, is divided into two sublayers. They are called the inner and the outer Helmholtz layer. The inner Helmholtz layer is the zone for specific adsorption of the traditional counterions, whereas the outer Helmholtz layer represents a zone with a gradual change into a diffuse layer. The Al ions in the figure can be replaced by Fe ions or any other metal ions. In essence, the triple-layer model has some similarities with a Stern double-layer model and, according to the latter authors, a triple layer will not develop if adsorption of potential-determining ions is absent. Here the Stern double-layer theory must be applied.

Effect of Electrolytes on the Thickness of the Diffuse Double Layer

The thickness of the diffuse double layer is dependent on the electrolyte concentration of the bulk solution (Table 5.7). High concentrations of electrolyte will usually result in suppression of the double layer. By increasing the electrolyte concentration, the amount of cations is increased. The latter reduces the concentration gradient in the liquid interface between the colloidal surface and surrounding liquid phase. Therefore, the tendency of cations to diffuse away from the surface of the colloids decreases, bringing about a decrease in thickness of the double layer.

Table 5.7 Effect of Concentration and Valences of Ions on Thickness of the Diffuse Double Layer

Electrolyte concentration (mol/L)	Thickness of diffuse double layer ($1/K$; cm)	
	Monovalent ions	Divalent ions
1×10^{-5}	1×10^{-5}	0.5×10^{-5}
1×10^{-3}	1×10^{-6}	0.5×10^{-6}
1×10^{-1}	1×10^{-7}	0.5×10^{-7}

Source: Verwey and Overbeek (1948).

Effect of Valency of Cations on the Thickness of the Diffuse Double Layer

The thickness of the diffuse double layer is also affected by the valency of the exchangeable cations. Generally, it has been reported that at equivalent electrolyte concentrations, monovalent cations in the exchange position yield thicker diffuse double layers than divalent cations (see Table 5.7). Trivalent cations will decrease the thickness of the double layer more strongly than divalent ions. This phenomenon is due to the tendency of ions to diffuse away from the colloidal surface (dissociate) in the following decreasing order:

Monovalent ions > divalent ions > trivalent ions

For example, Na and K ions are frequently reported to be responsible for relatively thicker doubler layers than Ca and Mg, whereas double layers formed by Al ions are comparatively the thinnest.

The Zeta (ζ) Potential

When a colloidal suspension is placed in an electric field, the colloidal particles move in one direction (toward the positive pole). The counterions move into another direction (toward the negative pole). The electric potential developed at the solid–liquid interface is called the *zeta (ζ) potential*. The seat of the ζ potential is the shearing plane or slipping plane between the bulk liquid and an envelope of water moving with the particle. Since the position of the shearing plane is unknown, the ζ potential represents the electric potential at an unknown distance from the colloidal surface. Van Olphen (1977) stated that the ζ potential is not equal to the surface potential. It is less than the electrochemical potential on the colloid. Perhaps, it is comparable with the Stern potential.

Effect of Electrolytes on Zeta Potential

The thickness of the double layer affects the magnitude of the ζ potential. Increasing the electrolyte concentration in the solution usually results in decreasing the thickness of the double layer. Compression of the double layer will also occur by increasing the valence of the ions in the solution.

The ζ potential may, therefore, be expected to decrease with increasing electrolyte concentration. It reaches a critical value at the point at which the ζ potential equals zero. This point is called the *isoelectric point*. At the isoelectric point, the double layer is very thin and particle repulsive forces are at a minimum. At and below this point repulsion would no longer be strong enough to prevent flocculation of colloidal particles. The ζ potential is not a unique property of the colloid, but depends on the surface potential (ψ) of the clay particle. It is determined from the electrophoretic mobility of the suspension using the formula

$$V_e = \frac{D\zeta E}{4\pi\eta}$$

where

$\quad V_e$ = electrokinetic velocity
$\quad D$ = dielectric constant
$\quad E$ = applied emf
$\quad \eta$ = viscosity of the fluid

The ζ potential in fact, is the electrokinetic potential at the slipping plane surface:

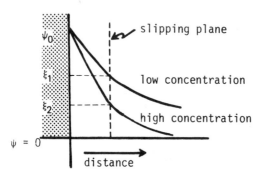

The surface potential of the colloid is ψ_0. In dilute solution, the electrokinetic potential has a value represented by ζ_1. By adding salt to the solution, the diffuse layer is suppressed and more counterions are forced to the colloid surface within the slipping plane. The slipping plane is considered to be at a fixed distance. Hence, at high salt concentration a change occurs in the total potential distribution as related to distance from the colloid. The potential distribution at high salt concentration is represented by the bottom curve in the figure. The electrokinetic potential ζ_2 is, therefore, smaller than ζ_1.

5.11 THE ELECTRIC DOUBLE LAYER AND STABILITY OF CLAYS

As discussed previously, clays carry a negative charge, which is ordinarily balanced by exchangeable cations adsorbed on their surface. In suspension, the cations tend to diffuse away from the clay surface into the bulk solution to balance the concentration difference occurring between the interface and bulk liquid phase. However, a large portion of these ions, especially those in the immediate vicinity of the clay surface, cannot move very far away because of the strong attraction from the negative charge on the clay surface. The cations aggregate in the interface, thereby forming an electric double layer, which may vary in thickness from 50 to 300 Å.

Whenever such clay particles approach each other, repulsion between the particles occurs because the outer parts of the double layers have the same type of charge (positive). The suspension is then considered stable, and the clay is considered to be dispersed. Because of this approach, the diffuse counterion atmospheres of the two particles interfere with each other. This leads to a rearrangement of the ion distribution in the double layers of both particles. Work must be performed to bring about these changes. The amount of work to bring about the changes is called *repulsive energy* or *repulsive potential* V_r at the given distance. The range and effectiveness of the repulsive potential depend on the thickness of the double layer. The repulsive force decreases, usually exponentially, with increasing distance between the particles (Figure 5.31). Opposite the repulsive forces, the clay suspension is also subjected to interparticle attraction. These forces of attraction are usually called the *van der Waals attraction* (V_A). The van der Waals attraction is only effective at very close distances, and decays rapidly with distance. However, since it is additive between atom pairs, the total attraction between particles containing many atoms is equal to the sum of all attractive forces between every atom of one particle and every atom of the other particle.

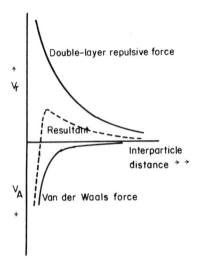

Figure 5.31 Double-layer repulsive and interparticle attractive (van der Waals) forces as a function of interparticle distance.

When the interparticle distance decreases to about 20 Å or less, van der Waals forces will become dominant, and the clay particles will flocculate. At interparticle distance of >20 Å, repulsive forces are dominant, creating a stable clay suspension. An example of the resultant of repulsive and van der Waals forces is shown in Figure 5.31 by the broken line curve.

Repulsion will dominate at low electrolyte concentration. The clay particles are shielded by relatively thick double layers, decreasing the possibility of mutual approach. At high electrolyte concentration, the chances of close approach are made possible by compression of the double layers. In this condition, van der Waals attraction may overcome the repulsive forces and coagulation or flocculation of colloidal particles occurs rapidly.

5.12 THE EFFECT OF FLOCCULATION AND DISPERSION ON PLANT GROWTH

The problem of flocculation and stability of soil suspensions is very important in soils. Stable aggregates can be formed only in soils containing clay that will flocculate. If clays remain dispersed, the soil is puddled. Puddled soils are sticky when wet and hard when dry. Root growth and soil aeration require a porous condition in soils. If percolating rainwater leaches out electrolytes from the soil, clay particles may become dispersed. As the soil becomes dry, caking or soil compaction may occur. The latter reduces the pore spaces, which inhibit soil aeration necessary for adequate root growth. Therefore, a flocculating concentration of electrolytes should be maintained in the soil. To reach such a condition, the soil should be limed, although acidic soils high in Al are usually flocculated. Calcium and Mg are known to have high flocculation powers on the negative clay particles and will reduce the toxic effect of high Al concentrations.

6

Adsorption in Soils

The electrochemical properties discussed in the preceding sections find many practical applications in soils. Besides the effect of flocculation on soil condition and plant growth, they are why soils develop the capacity to adsorb gas, liquid, and solid constituents. Cation exchange reactions, interactions between clay and organic compounds, complex reactions between metal ions and inorganic and organic colloids are additional implications of the electrochemical behavior of soil colloids.

6.1 TYPES OF ADSORPTION

The process of concentrating materials at the interface is called *adsorption*. It is one of the reactions attributed to the surface chemistry of soil colloids. Materials that decrease surface energy will be concentrated at the liquid–vapor interface, whereas materials that decrease interfacial energy will concentrate more in the liquid–solid interface (Gortner and Gortner, 1949). Adsorption in soils is more the type of concentrating material at solid–liquid interfaces. This type of adsorption can be distinguished into positive and negative adsorption. *Positive adsorption* is the concentration of the solute on the colloidal surfaces. It is also referred to as *specific adsorption*. The solute usually decreases surface tension. On the other hand, *negative adsorption* is the concentration of the solvent on the clay surface. The solute is then concentrated in the bulk solution; here, surface tension is increased.

6.2 ADSORPTION CHARACTERISTICS

Adsorption is dependent not only on the surface charge, but also on the surface area. The amount of material adsorbed is directly proportional to the specific surface. Adsorption is small, if the surface area is small, and increases with increased surface areas. Adsorption reactions are *reversible* and are *equilibrium reactions*. Sometimes an adsorption process results in chemical changes of the adsorbed material. The changes are of such a nature that desorption is inhibited; hence, the process is neither reversible nor in equilibrium. This type of adsorption is called *pseudo-adsorption*.

Adsorption is characterized by a *positive heat of adsorption*, meaning that energy is released during the adsorption process. The amount of heat produced because of adsorption of gas by solid surfaces can be expressed by the following equation (Gortner and Gortner, 1949):

$$\log h = \log a + b \log x$$

in which h = total heat released, x = amount of gas adsorbed in cubic centimeters (cm^3), and a and b are constants. This type of equation represents a linear regression, for which $\log a$ = intersect, and b = slope of the regression line, indicating rate (kinetics) of the adsorption process.

Adsorption generally decreases as temperature increases; in other words, adsorption is less at elevated temperatures. This is caused by an increased kinetic energy of the molecules at higher temperature, which interferes with the concentrating process. In contrast, the rate of a real chemical reaction increases as temperature is increased. Therefore, these differences can be used to distinguish an adsorption process from a true chemical reaction, although a similar equilibrium can be reached.

6.3 FORCES OF ADSORPTION

Forces responsible for adsorption reactions include (1) physical forces, (2) hydrogen bonding, (3) hydrophobic bonding, (4) electrostatic bonding, (4) coordination reactions, and (5) ligand exchange.

Physical Forces

The most important physical force is the van der Waals force. This force is a result of short-range dipole–dipole interactions. Its role is of importance only at close distances, since this type of force decreases rapidly with distance.

Hydrogen Bonding

As defined earlier, the bond by which a hydrogen atom acts as the connecting linkage is called a *hydrogen bond*. Water, which is dipolar, may become adsorbed at the clay surface through its linkage with hydrogen bonding.

Hydrophobic Bonding

Hydrophobic bonding is associated with adsorption of nonpolar compounds. The compounds compete with water molecules for adsorption sites and, in the process, adsorbed water is exchanged or expelled by the substance. Polysaccharides, for example, can be adsorbed in this way. The expulsion of water from clay surfaces, especially from intermicellar spaces, reduces swelling.

Electrostatic Bonding

Another type of force in adsorption is electrostatic attraction of substances, which is the result of the electrical charge on the colloid surface. This is the reason for (1) adsorption of water; (2) adsorption of cations, which leads to cation exchange reactions; and (3) adsorption of organic compounds. This may develop into complex reactions. An exchange of the organic ligand for a cation is called *ligand exchange*.

Coordination Reaction

The reaction involves coordinate covalent bonding. The latter occurs when the ligand donates electron pairs to a metal ion, usually a *transition* metal. The compound formed is called a *coordination compound, complex compound,* or *an organometal complex.* When it involves a reaction between an organic ligand and a metal ion only, the distinction between adsorbate and adsorbents may become obscured.

Ligand Exchange

Ligand exchange entails the replacement of a ligand by an adsorbate molecule. The adsorbate must have a stronger chelation capacity than the ligand. An example is presented in the foregoing paragraph on electrostatic bonding in relation to replacement of adsorbed organic compounds by cations. The replacement is not limited to expulsion by cations, but the adsorbed organic ligand can also be replaced by other organic substances.

6.4 ADSORPTION ISOTHERMS

Adsorption was defined earlier as the concentration of constituents at the colloidal surfaces. The curve relating the concentrations of adsorbed materials at a fixed temperature is called the *adsorption isotherm.* Four major types of equations are used to describe adsorption isotherms: the Freundlich equation, the Langmuir equation, the BET (Brunauer, Emmett, and Teller) equation, and the Gibbs equation.

Freundlich Equation

The adsorption isotherm in many dilute solutions is formulated by Freundlich (1926) as

$$x/m = kC^{1/n} \tag{6.1}$$

where

x = amount of material adsorbed

m = amount of adsorbents

C = concentration of the equilibrium solution

k, n = are constants

The value of $1/n$ is usually between 0.2 and 0.7 (Kruyt, 1944). For many pesticides at dilute concentrations $1/n = 1.0$ (R. A. Leonard, personal communication). The equation has no theoretical foundation and is empirical. The curve according to Eq. (6.1) is usually parabolic, and exhibits the following characteristic features: (1) there is no single point indicating that the process is completed; and (2) there is no region of discontinuity. By taking the logarithm, Eq. (6.1) changes into

$$\log x/m = \log k + 1/n \log C$$

The log equation gives a "straight-line" curve. Log k is the intercept, and $1/n$ represents the slope of the curve or the regression coefficient.

Another version of the Freundlich equation is

$$S = k_d C^n \qquad (6.2)$$

where

S = amount of solute retained per unit mass of soil (mg/kg)

k_d = distribution coefficient

C = solute concentration (mg/kg)

n = dimensionless parameter. Typically $n = 1$.

The distribution coefficient describes the partitioning of the solute between solid and liquid phases, and is considered analogous to the equilibrium constant in the mass action law equation. Strongly adsorbed heavy metals, such as Cu, Hg, Pb, and V exhibit high k_d values. The k_d value is also affected by pH and the cation exchange capacity (CEC). Soils with high pH and CEC values adsorb larger amounts of the heavy metals.

A third version of the Freundlich equation is called the *Van Bemmelen–Freundlich equation*:

$$\Gamma = Ac^\beta \tag{6.3}$$

where Γ = the amount adsorbed, c = the concentration of trace metal cations in the aqueous solution bathing the adsorbent, and A and β are empirical parameters, with $0 < \beta < 1.0$. Therefore, β is positive, but always lower than 1. This equation is used to describe adsorption of trace metals at constant temperature by clay minerals and sesquioxides in soils. It is an equation without chemical foundation, but it has the ability to describe adsorption data obtained at constant temperature (Sposito, 1980). The equation has often been associated with adsorption by heterogeneous surfaces. A homogeneous or uniform surface will adsorb ions according to the Langmuir equation, which is the topic of the next section.

Langmuir Equation

Another method to express adsorption is given by Langmuir (1916–1918):

$$\frac{x}{m} = \frac{k_1 C}{1 + k_2 C} \tag{6.4}$$

where

$$x = \text{amount adsorbed}$$
$$m = \text{amount of adsorbens}$$
$$k_1, k_2 = \text{are constants}$$
$$C = \text{concentration of equilibrium solution}$$

The difference from the Freundlich equation is the following. At very high concentration $k_2 C$ in Eq. (6.4) reaches such a value that the factor 1 can be neglected, so that the formula changes into

$$\frac{x}{m} = \frac{k_1}{k_2}$$

The latter formula stated that x/m becomes constant at high concentration. In other words, at high values of C, the surface of the adsorbens becomes saturated, and adsorption reaches a maximum.

Brunauer, Emmett, and Teller Equation

The BET equation was developed by these authors in 1938 for the adsorption of multilayers of nonpolar gases. The equation at low pressure is as follows:

$$\frac{P}{V(P_0 - P)} = \frac{1}{V_m C} + \frac{C - 1}{V_m C}\frac{P}{P_0} \tag{6.5}$$

where

P = equilibrium vapor pressure

P_0 = saturation vapor pressure

V = volume of gas adsorbed

V_m = volume of gas adsorbed when solid is covered with monolayer

C = constant related to heat of adsorption

The BET equation is an extension of the Langmuir for application to multilayer adsorption. It is assumed that the first layer of gas is attracted firmly to the surface, perhaps by van der Waals forces. The second and subsequent layers are held by weaker forces. As P/P_0 increases, the layers of gas are building up in an unrestricted way. The number of layers becomes infinite when P/P_0 is unity. If $P/[V(P_0 - P)]$ is plotted against P/P_0 (as the abscissa), a straight-line curve should be obtained. The slope is characterized by the factor $(C - 1)/V_m C$, and the intercept is at $1/V_m C$.

Gibbs Equation

The Gibbs equation describes adsorption processes in relation to surface tension:

$$\Gamma = -\frac{a}{RT}\left(\frac{\partial \gamma}{\partial a}\right)_T \tag{6.6}$$

where

Γ = surface concentration of adsorbed material

a = activity of solute in moles

R = gas constant

T = absolute temperature

γ = surface tension in dyn/cm

The solute is adsorbed on the surface of the adsorbens if $\partial\gamma/\partial a$ is negative. If $\partial\gamma/\partial a$ is positive, the solute is more concentrated in the bulk solution than in the interface region. The latter is sometimes called *negative adsorption*. The Gibbs equation finds primary application in adsorption at liquid–gas interfaces.

6.5 ADSORPTION OF WATER

Water is held in the soil in the pore spaces by forces of attraction at the colloid surfaces, by surface tension in the capillaries, and by attraction to the ions. In a wet condition, all the pores of the soil are filled with water. The soil is considered to be at its "maximum retentive capacity." In this condition, the matrix potential ψ_m is 0. The excess water is free to move in the soil by gravity. The movement, called *drainage*, usually results in several undesirable effects, such as waterlogging and leaching of nutrients.

As soon as excess water has drained away by gravity, water occurs in the macropores as thin layers (films) on the surfaces of soil particles and as wedges at the points of contacts between the particles. The micropores, on the other hand, are still filled with water. Such a condition is called *field capacity*. The force holding this water is approximately 0.2–0.3 bars. Living plants must satisfy their water need from field capacity water, by applying suction forces that can overcome the forces of adhesion, cohesion, and osmosis. As the water is taken up by the roots, the water films on soil particle surfaces become thinner. Hence, the forces of adsorption and retention become larger and larger. A point will eventually be reached at which the attraction of water by soil particles becomes so large that the plants can no longer extract enough to satisfy their needs. This point is called the *permanent wilting point*. The amount of water at this point is called the *permanent wilting percentage*. The force holding this water to the soil particles is estimated to be 15 bars.

Adsorption of Water by Silicate Clays

The adsorption of water by silicate clays is attributed to electrical forces. Upon adsorption, orientation of water takes place by the electrical field on the surface of clays, and the water molecules lose some of their freedom of movement. In terms of thermodynamics, it is said that the free energy of water has been decreased upon adsorption. This part of the energy is lost in the form of heat, as noticed by an increase in temperature at the wetting front as water reacts with soil. According to the BET adsorption theory, the heat of adsorption of the first layer of water is given by the constant C in Eq. (6.5). C itself can be found by the following relation:

$$C = a \exp \frac{Q_1 - Q_L}{RT}$$

where

  a = proportionality constant
  Q_1 = heat of the first adsorbed layer of water
  Q_L = heat of liquefaction (becoming liquid)
  R = gas constant
  T = absolute temperature

For practical purposes, $a = 1$ (Taylor and Ashcroft, 1972) and the equation can be changed to

$$2.303 \log C = \frac{Q_1 - Q_L}{RT}$$

The heat of adsorption, or the difference in energy of adsorption at the interface, $Q_1 - Q_L$, varies with the type of clay minerals. Taylor and Ashcroft (1972) reported that it was low for illites and other 2:1 types of minerals, but that it was comparatively higher for kaolinite and other 1:1 types of clays.

  Clays with large internal surfaces are also capable of adsorbing water in the intermicellar regions, and interlayer hydrates can be formed. The degree of stability and the structure and characteristics of interlayer water are dependent on the presence of interlayer cations and on the composition of the interlayer clay surfaces. Interlayer water reacts both with the oxygens of the siloxane surface and with cations present in the intermicellar spaces. In the event that only one to three layers of water are adsorbed, some degree of water stability has been noted. Because of the strong bonding of this water, the behavior of the adsorbed interlayer water is relatively static. On the other hand, when more water layers are adsorbed in internal spaces, this water behaves more liquidlike, and a higher degree of mobility is then exhibited. Interlayer water can form a link between cations and the negatively charged clay surface. They also are capable of donating protons to bases in interlayer positions and can donate or accept electrons from two neighboring ions. Proton transfer from water to bases often takes place more readily in the intermicellar spaces than in the free soil solution. Formation of protonated bases in interlayer solutions is considered the result of increased acidity and ionization of interlayer water.

The adsorption of water in intermicellar spaces of expanding 2:1 lattice-type clays (e.g., montmorillonite) usually results in an increase in basal spacings, as determined by x-ray diffraction analysis (Table 6.1). The data in Table 6.1 seem to confirm that the amount of water present and nature of interlayer cation both have an influence on adsorption. However, increased humidity appears to expand the lattice space more than the type of cation.

Two types of interlayer water can be distinguished (Walker, 1961): (1) water in the first (inner) hydration shell coordinated

Table 6.1 Basal, d (001) Spacings (Å) of Montmorillonite Saturated with Different Cations, as Affected by Interlayering with Adsorbed Moisture

	Interlayer cations					
	Li^+	Na^+	K^+	Ca^{2+}	Mg^{2+}	Ba^{2+}
Increasing humidity	9.5	9.5	10.0	9.5	9.5	9.8
	15.4	15.4	15.0	15.4	15.4	15.5
	19.0	19.0		18.9	19.2	18.9
	22.0					

Source: Norrish (1954).

around the interlayer cation and (2) water in the secondary (outer) coordination sphere of the cation. The latter is more mobile than water in the first coordination sphere. The presence of these two types of interlayer water was discovered by x-ray analysis and infrared spectroscopy. With infrared analysis water in the two coordination spheres is distinguished by differences in intensity of infrared absorption at 3600–3400 and at 1620 cm^{-1}, for OH-stretching and OH-bending vibrations, respectively, of the water molecule. Water in the first coordination sphere has an infrared absorption at 3600 cm^{-1} and a very strong absorption at 1620 cm^{-1}, perhaps because of the weaker hydrogen bonding involved in the adsorption process. In contrast, water in the second coordination sphere, or the more mobile water, exhibits a spectrum similar to that of free water, with absorption bands between 3600 and 3400 cm^{-1} and a relatively weak band at 1620 cm^{-1}. The OH-stretching vibration of interlayer water between 3600 and 3400 cm^{-1} is usually difficult to observe, since structural OH groups of montmorillonite may also absorb at or near 3600 cm^{-1}.

Interlayer water can be displaced by other polar compounds. Generally, water adsorbed in the second coordination sphere is more easily displaceable than water held in the first coordination shell. Several polar compounds, such as pyridine and nitrobenzene, are capable of displacing the mobile interlayer water of the outer shell. Removal of water from the first coordination shell takes place only by treatments that effect dehydration of the cation. This displacement of water from the primary shell causes the cation to polarize the residual water, with the consequent dissociation of the water molecules. The protons produced by the dissociation reaction may be used for protonation of other organic compounds.

Water Adsorbed by Organic Matter

Organic matter is known to contain several functional groups, such as carboxylic groups, phenolic and alcoholic hydroxyl groups, amino acid groups, amides, ketones, and aldehydes. Among these many groups, perhaps the most important sites for adsorption of water are provided by the carboxyl groups. Ionized carboxylic groups possess high affinity for water, although the other functional groups may also exhibit some degree of adsorption.

Water can be held by a single hydrogen bond or by multiple bonds

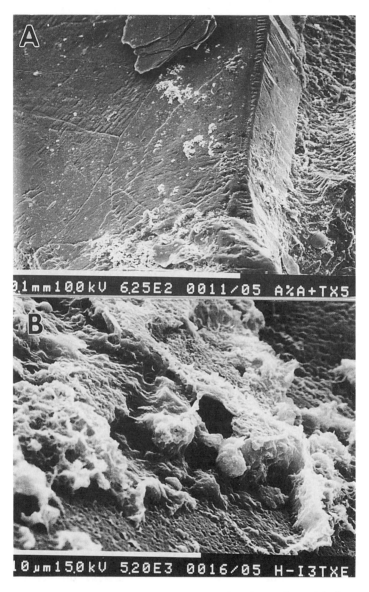

Figure 6.1 Scanning electron micrographs of a sand particle from a normal and localized dry spot in a golf green. (A) Normal or healthy area of a golf green. Note the absence of coatings on the sand particle surface. (B) Localized dry spot. The sand particle surface is covered by a fulvic acid coating. (Courtesy of K. J. Karnok, unpublished personal files, University of Georgia.)

or coordination bonds. Coordinated water is usually held more strongly by organic matter than is water held by a single bond. Upon drying and exposure of polar sites of organic matter, an internal pairing of proton donor groups (OH) and proton acceptor groups (C=O) may occur. These reactions can produce stable sites of pairing, so that the organic compound will not rehydrate upon wetting. The latter is possibly part of the reasons for the irreversible behavior of organic matter in rehydration upon wetting and drying.

This phenomenon of irreversible drying of organic matter causes many problems in a number of soils, and it is especially harmful for the establishment of golf greens (Tucker et al., 1990; Miller and Wilkinson, 1979). The acquired soil hydrophobicity is related to formation of fulvic acid coatings around soil particles (Figure 6.1). The dry fulvic acid coatings repel water and create the so-called *localized dry spots* in golf greens. It is believed that the fulvic acid coating is a metabolic product of soil microorganisms. Wilkinson and Miller (1978) have isolated fungi species of *Helminthosporium*, *Alternaria*, and *Culvularia* in humic matter from localized dry spots.

6.6 PLANT–SOIL–WATER ENERGY RELATION

The water potential of plant cells is defined as

$$\psi_{w(plant)} = \psi_p + \psi_s + \psi_m$$

When a plant root is immersed in the soil solution, water moves into or out of the plant cells, until equilibrium is reached. At equilibrium, the water potential of the plant cells equals the water potential of the soil solution:

$$\psi_{w(plant)} = \psi_{w(soil)}$$

Substituting this relation in the previous equation gives

$$\psi_{w(soil)} = \psi_p + \psi_o$$

where $\psi_o = \psi_s + \psi_m$ = osmotic potential

By rearranging the equation we find

$$\psi_p = \psi_{w(soil)} - \psi_o \tag{6.7}$$

The permanent wilting point occurs at $\psi_p = 0$ (Slatyer, 1957). Therefore

$$\psi_{w(\text{soil})} - \psi_o = 0$$

or

$$\psi_{w(\text{soil})} = \psi_o$$

If the soil–water potential decreases, soil–water content also decreases. The latter may cause water to move out of the cell, ultimately producing a phenomenon called *cytoplasmolysis*.

It is generally believed that at the *permanent wilting point* most plants will not be able to take up water and eventually die because of lack of water. However, some plants may respond differently to prolonged conditions of lack of water. In conditions of water stress, the plants need to conserve water, but at the same time they need to continue to produce photosynthates to maintain metabolic activities. With many plants, the level of metabolic activity is reduced. This can be accomplished by reducing their biomass through shedding their leaves, or by going into a dormant stage. Other plants, especially plants in semiarid and arid regions, respond differently. They have acquired a mechanism by which the uptake of CO_2 for photosynthesis can temporarily be disengaged during periods of severe water stress. The plants adapt to the harmful condition by regulating their stomatal openings to enable the entrance of small, but sufficient amounts of CO_2 for maintaining a minimum level of metabolic activity, but not so much that water loss becomes harmful (Fisher et al., 1981; Turner and Kramer, 1980). Therefore, plant life below the permanent wilting point is made possible by regulating stomatal aperture and other metabolic activities that are associated with changes in plant water potentials. As water content in plant tissue decreases, the osmotic potential is lowered owing to net accumulation of solutes. This lowering of the osmotic potential is called *osmotic adjustment*. It usually contributes to a partial or full maintenance of turgor pressure. The rate of change in osmotic potential, as a result of passive changes in solute concentration, can be expressed by the equation (Turner and Jones, 1980):

$$\psi_o = \frac{\phi_0 - V_0}{V}$$

in which ψ_o = osmotic potential, ϕ_0 = osmotic potential at full tur-gor or at zero turgor, V = osmotic volume, and V_0 = osmotic volume at full turgor or zero turgor. The equation is not valid for deter-mining osmotic adjustment owing to active accumulation of solutes in plant cells. Active accumulation of solutes will yield a decrease in osmotic potential far greater than can be predicted from the fore-going equation.

6.7 ADSORPTION OF ORGANIC COMPOUNDS

Adsorption of organic compounds by soil components has recently received increasing attention, especially adsorption of pesticides and herbicides. Physical, hydrogen, electrostatic, and coordination bond-ing are involved in the process (Bailey and White, 1970). Physical bonding, from the London–van der Waals attraction, is limited to external surface reactions only. Hydrogen bonding may occur if the compound has an N—H or OH group that can be linked to the O on the clay surface. Electrostatic bonding takes place between organic cations and anions and clays. The organic cations are mostly ad-sorbed on the negative clay surfaces, whereas the organic anions are more attracted toward the edge surfaces of clay (Van Olphen, 1977). Intermicellar adsorption between the clay layers is limited to low molecular weight organic compounds (Tan and McCreery, 1975). The presence of organic compounds in intermicellar positions results in an expansion of the basal spacings. For example, mont-morillonite treated with ethylene glycol or glycerol, gives a basal spacing of 17 Å, in contrast with 9.2–9.5 Å, its ordinary spacing in the presence of water or exchangeable cations. In the adsorption process, the organic molecule may displace adsorbed water from the clays. In turn, the adsorbed organic compound can often be removed by washing with water, except perhaps for those in the intermicellar spaces. Coordination bonding results in formation of complex com-pounds. It is formed by the organic ligand donating electron pairs to a metal or an ion. The metal takes the central position and is surrounded by a group of ligands. The metal ion, complexing organic matter, can also be attached to a clay particle. The metal then forms a bridge between the organic compound and the clay surface.

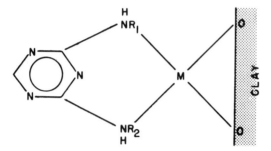

Uncharged organic compounds are also now known to be subject to adsorption by clay minerals. The adsorption can be direct or indirect, and physical forces (e.g., van der Waals forces) are mainly responsible for such an attraction. For the physical force to become effective, the uncharged organic compounds have to be relatively large in size or chain length. A maximum chain length of five units has been reported to effect adsorption significantly (Theng, 1974). Molecules of larger size, or with more than five units, can be adsorbed in the presence of excess water. The increase in adsorption with longer chain length is attributed to van der Waals forces becoming more effective as the size of molecules is increased.

Indirectly, uncharged organic compounds are adsorbed by the silicate surface through linkage to the exchangeable cations. Evidence for the latter came from infrared spectroscopy. Competition for ligand sites usually occurs between the organic compounds and water molecules around the exchangeable cations. The organic compound can either react directly with the cation, or it can be indirectly coordinated to the cation by linkage with water molecules in the hydration shell. This type of reaction, called *ion–dipole reaction*, depends on the polarizing power of the cation, the basicity of the organic compound, and on the nature of packing when adsorbed in intermicellar spaces.

Another adsorption mechanism reported in the literature is the adsorption caused by the presence of activated C—H groups. Activation of methylene groups by adjacent electron acceptors makes it possible for the activated methylene groups to form hydrogen bonds with the oxygens of the siloxane surface of clay minerals. However, in later studies many arguments have been presented on the possibility for formation of a C—H···O type of bonding.

Nature of Adsorption Isotherms

There is no general agreement in the literature on the nature of adsorption isotherms. Bailey and White (1970) are of the opinion that adsorption of organic matter can be better described with the Freundlich equation, but Weber (1970) has reported that adsorption of organic matter follows the Langmuir equation. According to the Freundlich equation, theoretically, adsorption increases indefinitely with increasing concentration. On the other hand, the Langmuir equation indicates that adsorption of organic matter on the clay surface tends to reach a maximum limit. The latter is compatible with the fact that soil and clay do not have an infinite capacity to absorb, but will sooner or later be saturated. Adsorption of organic

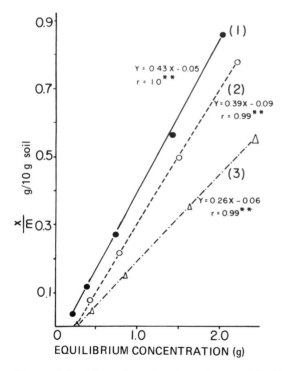

Figure 6.2 Adsorption of water extracts of broiler litter by Cecil topsoil at (1) 25°C, (2) 35°C, and (3) 50°C. (Adapted from Tan et al., 1975.)

compounds, such as humic acid and the like, following the Lang-
muir-type equation has also been reported by Inoue and Wada (1973)
and Tan et al. (1975). In their studies with poultry litter extracts
adsorbed by kaolinite and montmorillonite minerals, Tan et al.
(1975) show that the slope of the adsorption isotherm decreases with
increasing temperature from 25°, to 35°, to 50°C (Figure 6.2). They
also indicate that infrared analysis confirms that protonated compounds are adsorbed in larger amounts than Na^+-saturated extract.
The latter may suggest that ionic bonding has been involved in the
adsorption reaction.

Effect of Molecular Size on Adsorption

The size of the organic molecule is considered to play an important
role in its rate of adsorption. Bailey and White (1970) summarize
the effect of molecular size as follows:

1. The adsorption of nonelectrolytes by nonpolar adsorbents in-
    creases as molecular weights of the substances increases.
2. The van der Waals forces of adsorption increase with increas-
    ing molecular size.
3. The adsorption decreases because of steric hindrance.

The evidence currently available also shows the presence of a
maximum limit in molecular size in adsorption of organic compounds. As stated earlier, appreciable adsorption is noticed with
compounds having a chain length of five units, whereas larger molecules (chain length greater than five units) may be adsorbed only
in the presence of excess water. However, very large molecules can
experience difficulties in adsorption owing to adverse molecular configuration. Inoue and Wada (1973) succeeded in determining the
molecular size limit for adsorption. They reported that humic molecules with molecular weights between 1500 and 10,000 are pref-
erentially adsorbed over the smaller or larger molecules. These size
limits have been confirmed by Tan (1976a) in his studies with
Sephadex gel filtration. To study complex formation, humic acid was
shaken with kaolinite or montmorillonite, and the remaining humic
acid in solution was fractionated with Sephadex G-50 gel filtration.
The results (Figure 6.3) showed that mainly the fraction with a
molecular weight of 1500 (HA-II) was adsorbed. Before treatment

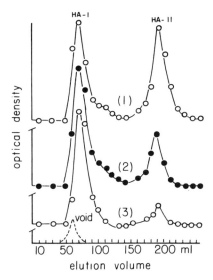

Figure 6.3 HA-I is high-molecular-weight humic acid (MW > 30,000) and HA-II is low-molecular-weight humic acid fractions (MW < 1500) separated by Sephadex G-50 gel filtration: (1) original humic acid, (2) humic acid after interaction with kaolinite, and (3) humic acid after interaction with montmorillonite. (From Tan, 1976a.)

with kaolinite or montmorillonite, the original humic acid was composed of equal amounts of high (HA-I: >30,000) and low molecular weight (HA-II: <1500) fractions. After interaction with the clay minerals, only the elution peak of the low-molecular-weight fraction was decreased in intensity, meaning that the equilibrium humic acid solution has lost some of the low-molecular-weight humic acid. The amounts of humic acid lost had apparently been adsorbed by kaolinite or montmorillonite. The elution curves also provide indications that montmorillonite has a larger adsorption capacity than kaolinite. The elution peak of HA-II was significantly smaller after treatment with montmorillonite than with kaolinite. This difference in adsorption owing to the type of clays had been expected. The expanding 2:1 minerals have properties, such as high CEC and surface area, that give rise to large coulombic and van der Waals forces. On the other hand, the nonexpanding 1:1 type of clays have low

CEC and low surface area, all properties that will not contribute to considerable adsorption.

Interlayer Adsorption and Molecular Orientation of Organic Compounds

Adsorption of organic compounds by clay surfaces can occur not only on the outer surfaces, but x-ray diffraction analysis of basal spacings of clays has shown that considerable amounts of the organic molecules can penetrate the intermicellar regions of 2:1 expanding types of clays especially. This placement of organic compounds in interlayer spaces of silicate minerals is called *intercalation* or *solvation*. Intercalation occurs more frequently with the 2:1 than with 1:1 types of minerals. Because of the strong bonds between the layers of kaolinite, penetration of intermicellar spaces by organic compounds is very difficult, although intercalation, in some instances, can be achieved with certain inorganic salts and polar substances. Therefore, adsorption of organic compounds are, in general, confined to the outer surfaces and to the edges of kaolinite, where the unsatisfied valencies by broken bonds have been estimated to amount to approximately 10–20% of the crystal area. In contrast with kaolinite, montmorillonite has both active outer and interlayer surfaces. Of the latter two, the interlayer surface accounts for the major portion of the total surface area. Only 10% of the total surface area is made up of active crystal edges, thereby reducing the importance of edge effects in adsorption.

Interlayer adsorption is affected by molecular size, polarity, and polarizability of the organic substances. In addition, it appears from previous discussions that water molecules and inorganic cations in intermicellar positions may also play an important role in adsorption. For the substance to be adsorbed on the intermicellar surfaces, sufficient energy must be available to exceed, or at least equal, the energy forces holding the layers together. The presence of polar molecules, such as water, may serve as an aid to the latter, since interlayer water is capable of keeping the layers apart, and in this way decreasing the electrical field between the clay surface-exchangeable ion. Evidence for this was obtained with benzene, which can be adsorbed by montmorillonite only when interlayer water is present (Bailey and White, 1970).

Polarizability is the ease with which negative and positive charges in a molecule can be displaced relative to one another in the presence of an electric field. From the definition, it is perhaps clear that the greater the ease of polarizability of the molecule, the greater will be its adsorption in intermicellar spaces.

X-ray diffraction analysis of expanding silicate clays solvated with ethylene glycol reveals that polarity and polarizability play an important role in orientation and spatial arrangement of interlayer organic molecules. Two kinds of orientation have been reported for aliphatic chain molecules adsorbed with their axis perpendicular to the silicate surface (Theng, 1974): (1) an interlayer arrangement in which the plane of carbon zig-zag is perpendicular to the silicate layer, and (2) an interlayer arrangement in which the zig-zag carbon plane is parallel to the silicate surface. Strongly polar compounds will favor orientation according to arrangement (2), whereas nonpolar compounds tend to arrange themselves according to orientation type (1).

6.8 INTERPARTICLE ATTRACTION

Interaction between charged colloidal particles occurs frequently in soils. As discussed previously, the presence of a double layer creates repulsive forces between particles. The colloids also possess London–van der Waals forces responsible for attraction of particles to each other. The latter can be significant at close range, but will be of no importance at large interparticle distances. The net balance between the repulsive and attractive forces determines interparticle attraction. For most soils, Al^{3+}, H^+, or Ca^{2+}, and Mg^{2+} ions are dominant in the double layers. These ions, except H^+, tend to reduce the repulsive forces of the double layer. The van der Waals forces will then be effective and the soil particles interact and form aggregates. In contrast with the foregoing ions, Na^+ ions have a different effect. Because of the large hydration shell, Na^+ ions tend to increase the double-layer repulsive forces. The soil particles are not aggregated and when wet remain in a dispersed condition. Soils in which Na^+ ions are dominant, such as the saline soils, may have a poor physical condition.

7

Cation Exchange

7.1 ADSORPTION OF CATIONS BY SOIL COLLOIDS

Since clay colloids carry negative charges, cations are attracted to the clay particles. This is Mother Nature's condition to maintain electroneutrality in soils. These cations are held electrostatically on the surface of the clay. Most of them are free to distribute themselves through the liquid phase by diffusion. The density of ion population is greatest at or near the surface. These cations are called *adsorbed cations*. Different rates and orders of adsorption are known among the cations, since the adsorption reaction depends on the surface potential, valence, and hydrodynamic radius. As surface potential increases, more of the divalent ions will be adsorbed. Trivalent cations would even be more strongly concentrated on the clay surface exhibiting high surface potential values. When a mixture of monovalent and divalent ions are present in a soil solution, adsorption is usually shifted in favor of the divalent ions. Specific adsorption of cations is also affected by the hydrodynamic radius. The crystalline radius, on the other hand, plays only a minor role. Generally, ions with smaller hydrated sizes are preferably adsorbed. The following decreasing order of preference for adsorption of monovalent cations by clays has been reported (Gast, 1977):

$Cs > Rb > K > Na > Li$

The rate of adsorption for Cs is the highest, because Cs is the smallest in hydrated size. The ion has a thin hydration shell, which makes a close approach to the clay surface possible. Lithium, on the other hand, has the largest hydrodynamic radius. Its thick hydration shell increases the distance from the ion to the clay surface. Such a series of ions, listed in decreasing order of preferential adsorption, is called a *lyotropic series*. Lytropic series for divalent cations has also been reported in the literature (Taylor and Ashcroft, 1972):

$Th > La > Ba > Sr > Ca > Mg$

Evidence is present that different lyotropic series exist for different types of clays.

7.2 CATION EXCHANGE REACTIONS

The term *cation exchange* is preferred over the term *base exchange*, since the reaction also involves H^+ ions. The hydrogen ion is a cation, but not a base. The adsorbed cations can be exchanged by other cations, hence the cations are also called *exchangeable cations*. The process of replacement is called *cation exchange*. The rate of reaction is virtually instantaneous. To maintain electroneutrality in the soil, exchange reactions are stoichiometric reactions, as illustrated by the classic experiment of Way (1850):

$Ca-soil + 2NH_4^+ \rightarrow (NH_4)_2-soil + Ca^{2+}$

Adsorption and cation exchange are of great practical significance in nutrient uptake by plants, soil fertility, nutrient retention, and fertilizer application. Adsorbed cations are generally available to plants by exchange with H^+ ions generated by the respiration of plant roots. Nutrients added to the soil, in the form of fertilizers, will be retained by the colloidal surfaces and are temporarily prevented from leaching. Cations that may pollute the groundwater may be filtered by the adsorptive action of the soil colloids. As such, the adsorption complex is considered to give to the soil a storage and buffering capacity for cations. In addition, it may play a role in making liming materials available to plant growth. Calcitic limestone, or $CaCO_3$, is insoluble in water. When added to acidic soil (Al–soil), limestone may react with H_2O containing CO_2:

$$CaCO_3 + H_2CO_3 \rightarrow Ca(HCO_3)_2$$

The calcium bicarbonate formed is soluble in water. Calcium that is dissociated off can then be adsorbed by the soil in exchange for Al^{3+}:

$$\tfrac{3}{2}Ca(HCO_3)_2 + Al\text{–soil} \rightleftharpoons (Ca)^{3/2}\text{—soil} + Al(OH)_3 + 3CO_2$$

Thomas (1974) considered these processes a neutralization and precipitation type of cation exchange reaction.

7.3 CATION EXCHANGE CAPACITY

The *cation exchange capacity* (CEC) of soils is defined as the capacity of soils to adsorb and exchange cations. Scientifically, it is related to the surface area and surface charge of the clay. This relation is expressed by the equation

$$CEC = S \times \sigma \tag{7.1}$$

where

S = specific surface

σ = surface charge density

The surface charge density can be calculated using the equation

$$\sigma = \sqrt{2\epsilon nkT/\pi} \; \sinh(ze\psi_0/2kT) \tag{7.2}$$

where

ϵ = dielectric constant

n = electrolyte concentrations in number of ions per milliliter

k = Boltzmann constant

T = absolute temperature (K)

z = valence

e = electron charge in esu

ψ_0 = surface potential in Stat volts

However, CEC is commonly determined by extraction of the cations from soils with a solution containing a known cation for exchange. The results, expressed in milliequivalents per 100 g of soils (= cmol $(+)$ kg^{-1}), are taken as the CEC of soils. The U.S.D.A. Soil Survey Division uses as a unit milliequivalents per 100 g of clay. The values for CEC may vary considerably, depending on the (1) concept used of CEC, (2) methods of analysis, (3) type of colloids, and (4) amount of colloids. It is common practice in the determination of CEC to analyze all exchangeable cations. The CEC is then

$$CEC = \sum mEq \text{ exchangeable cations per 100 g soil}$$

Bolt et al. (1976) are of the opinion that a certain correction is needed to this procedure. They stated that the real adsorbed cations are not accompanied by anions. But "free" cations may have slipped in carrying with them their counterions as "stowaways." These ions may have been analyzed together with the "true" exchangeable cations. Hence, the ions of these free salts must be subtracted to obtain the correct CEC.

Example: Determination of exchangeable cations and anions gives the following results:

Cations (mEq/100 g)	Anions (mEq/100 g)
Na^+ = 5	Cl^- = 0.8
K^+ = 5	HCO_3^- = 0.2
Ca^{2+} = 10	$\sum$ anions = 1.0
Mg^{2+} = 10	
H^+ = 5	
$\sum$ cations = 35	

The CEC, according to Bolt et al. (1976) is

$$CEC = \sum \text{exchangeable cations} - \sum \text{exchangeable anions}$$
$$= 35 - 1.0 = 34 \text{ mEq/100 g}$$

The value for soil CEC varies according to type and amounts of colloids present in soils. On the average the CEC of the major soil colloids is as follows:

Soil colloids	CEC (mEq/100 g)
Humus	200
Vermiculite	100–150
Montmorillonite	70–95
Illite	10–40
Kaolinite	3–15
Sesquioxides	2–4

Another method to determine CEC is calculation from the formula weight of the particular clay mineral. Here, the charge per formula weight (FW) should be known (see Table 5.2) or first determined. An example of a calculation is given for smectite

Mineral	Formula	Charge/FW
Smectite	$(Si_8)(Al_{3.33}Mg_{0.67})O_{20}(OH)_4nH_2O$ $\mid$ $Na_{0.67}$	0.50 Eq

$$FW \text{ of smectite} = (8 \times 28.1) + (3.33 \times 27) + (0.67 \times 24.3)$$
$$+ (0.67 \times 23) + (20 \times 16) + (4 \times 16) + (4 \times 1)$$
$$= 738\,g$$
$$CEC \text{ of smectite} = 0.50/738$$
$$= 0.0006775\,Eq/g \quad or \quad 67.8\,mEq/100\,g$$

Depending on the types of charges, several types of CECs can be distinguished:

1. CEC_p: This is the CEC attributed to permanent charges, and is of importance especially in 2:1 layer types of minerals.

2. CEC_t: This is the total CEC, or the CEC produced by both permanent and variable negative charges.

3. CEC_v: This is called the variable charge CEC, or the CEC caused by the variable negative charges. It is in fact CEC_t − CEC_p. CEC_v is common in soils high in organic matter content, sesquioxides, amorphous, or 1:1 layer types of clays.

7.4 THE EXCHANGING POWERS OF CATIONS

Different cations may have different abilities to exchange adsorbed cations. The amount adsorbed is often not equivalent to the amount exchanged. Divalent ions are usually held more strongly than monovalent ions. They will be exchanged with more difficulty. It is sometimes noticed that if Ba^{2+} is used as the exchange cation, the exchange does not occur in equivalent amounts. Barium is strongly adsorbed by the clay, but appears to have low-penetrating power. Therefore, it exchanges less than would be expected from the amount of Ba adsorbed. On the other hand, cation exchange using NH_4^+ ions may often yield higher exchange results than is expected from amounts of NH_4^+ ions adsorbed. Ammonium as a monovalent ion will be attracted less strongly than the Ba ions, but NH_4^+ has high-penetrating power.

An exception to this is perhaps the use of H^+ ions. Hydrogen ions are adsorbed more strongly than the other monovalent or divalent ions. Hydrogen clays prepared by exchange reactions initially contain large amounts of exchangeable H^+ ions and small amounts of Al^{3+} ions. However, the concentration of exchangeable Al^{3+} ions builds up rapidly. The exchangeable H^+ ions cause a partial decomposition, and Al released from the clay becomes exchangeable.

7.5 THE IONIC COMPOSITION OF THE EXCHANGE COMPLEX

The soil solution contains a mixture of cations. All of them are subject to attraction to clay surfaces. In the preceding sections, we have seen that the adsorption and exchange of cations depend on the concentration and the nature of replacing (added) cations. The nature and the composition of the soil solution will change accordingly, depending on the type and amounts of cations being replaced and on the concentration of the end products, as determined by solubility and dissociation. Understanding the changes that can occur in ionic composition is very important in practice. We can then manipulate this composition through fertilizers and lime addition in favor of plant growth. To be able to study this problem, means must be available to formulate the ionic composition in the soil in the presence of charged colloids. Several equations have been reported for this purpose.

7.6 EMPIRICAL EQUATIONS OF CATION EXCHANGE

The Freundlich Equation

The adsorption equation of Freundlich is one method to express ionic composition in the soil solution. It is adaptable to adsorption reactions in a narrow range. The equation is as follows:

$$x = kC^{1/n} \tag{7.3}$$

where

x = amount of cations adsorbed per unit amount of adsorbent

C = equilibrium concentration of the added cation

k, n = constants

The Freundlich equations is not exactly a formulation of cation exchange reactions. These pertain to formulations of adsorption of cations, and no exchange parameter is included in the equation.

The Langmuir–Vageler Equation

$$\frac{x}{x^\circ} = \frac{kC}{1 + kC} \tag{7.4}$$

where

x = amount of cations adsorbed per unit weight of exchanger

x° = total exchange capacity

C = concentration of added cations in moles per liter

k = affinity coefficient

The constant k can be determined as follows (Thomas, 1974). Rearranging Eq. (7.4) gives

$$x(1 + kC) = x°kC$$
$$x = xkC = x°kC$$
$$x = x°kC - xkC$$
$$x = kC(x° - x)$$
$$kC = \frac{x}{x° - x}$$

therefore

$$k = \frac{x}{C(x° - x)} \tag{7.5}$$

7.7 MASS ACTION LAW EQUATIONS OF CATION EXCHANGE

Kerr's Equation

The equation can apply for a (1) mono–monovalent ion exchange or (2) mono–divalent cation exchange reaction.

Mono–Monovalent Cation Exchange Reaction

Assume that we have the following exchange reaction:

$$Na^+ + K\text{–soil} \rightleftharpoons Na\text{–soil} + K^+$$

Application of the mass action law gives

$$\frac{[Na^+](K^+)}{(Na^+)[K^+]} = k_{eq} \tag{7.6}$$

or

$$\frac{(K^+)}{(Na^+)} = k_{ex} \frac{[K^+]}{[Na^+]}$$

The sign [ ] denotes adsorbed ions, whereas ( ) denotes free ions in solution. The equilibrium constant k_{eq} becomes k_{ex}, which is often called the *selectivity coefficient*. The value of k_{ex} indicates the tendency that one cation is adsorbed more over the other. When

$$k_{ex} = \frac{(K^+)}{(Na^+)}$$

then:

$$\frac{[K^+]}{[Na^+]} = 1.0$$

This means that equal amounts of K^+ and Na^+ ions are adsorbed.

Mono–Divalent Cation Exchange Reaction

Assume the exchange reaction:

$$2Na^+ + Ca-soil \rightleftharpoons (Na)_2-soil + Ca^{2-}$$

Kerr indicates that at equilibrium, the reaction can be described by the following equation

$$\frac{[Na^+](Ca^{2+})}{[Ca^{2+}](Na^+)^2} = k \tag{7.7}$$

The signs [ ] and ( ) again denote adsorbed and free ions, respectively. Conforming to the mass action law, the equation says that the ratio of the activity product of reaction products and that of the reactants is constant. By taking the square root, Eq. (7.7) changes into

$$\frac{[Na^+](\sqrt{Ca^{2+}})}{[\sqrt{Ca^{2+}}](Na^+)} = k \tag{7.8}$$

This equation is also known as the *Gapon equation*.

Vanselow's Equation

Vanselow considers the activity of adsorbed cations proportional to the mole fraction of total occupying cations present:

$$\text{Proportion of Na adsorbed} = \frac{[Na^+]}{[Na^+ + Ca^{2+}]}$$

$$\text{Proportion of Ca adsorbed} = \frac{[Ca^{2+}]}{[Na^+ + Ca^{2+}]}$$

By substituting the above in Eq. (7.7) gives

$$\frac{([Na^+]/[Na^+ + Ca^{2+}])^2(Ca^{2+})}{([Ca^{2+}]/[Na + Ca^{2+}])(Na^+)^2} = k$$

Rearranging this gives Vanselow's equation:

$$\frac{[Na^+]^2(Ca^{2+})}{[Ca^{2+}][Na^+ + Ca^{2+}](Na^+)^2} = k \tag{7.9}$$

Again [ ] denotes adsorbed ions, and ( ) denotes ions in solution. Krishnamoorty and Overstreet (1949) have developed a similar equation.

7.8   KINETIC EQUATIONS OF CATION EXCHANGE

The kinetic type of formulation has been developed by Gapon (1933), Jenny (1936), and by Davis (1945). They are in essence the same formula as expressed by Eq. (7.8).

7.9   EQUATIONS BASED ON THE DONNAN THEORY

A Donnan system is a system composed of solution i and o, separated by a semipermeable membrane (i = inside solution, o = outside solution):

Solution i	Solution o
Na^+	Na^+
Cl^-	Cl^-
Na-clay	

Semipermeable
membrane

Solution i contains Na^+ and Cl^- and Na–clay, whereas solution o contains only Na^+ and Cl^- ions of different concentrations from those in solution i. The membrane is permeable only to Na^+ and Cl^- ions; therefore, only these ions will move and distribute themselves in solutions i and o until equilibrium is reached. At equilibrium the following relation holds:

$$(Na^+)_i(Cl^-)_i = (Na^+)_o(Cl^-)_o$$

or

$$\frac{(Na^+)_i}{(Na^+)_o} = \frac{(Cl^-)_o}{(Cl^-)_i}$$

Donnan systems are present in soils and are of special importance in soil solution–plant root relationships. They have been extended to cation exchange phenomena and predict essentially the same as the mass action law:

$$\frac{[Na]^2(Ca^{2+})}{(Na^+)^2[Ca^{2+}]} = k$$

Donnan assumes $k = 1$; therefore, the above equation changes into

$$\frac{[Na^+]^2(Ca^{2+})}{(Na^+)^2[Ca^{2+}]} = 1$$

or

$$\frac{[Na^+]}{(Na^+)} = \frac{[\sqrt{Ca^{2+}}]}{(\sqrt{Ca^{2+}})} \tag{7.10}$$

7.10 EQUATION OF ERIKSSON

Eriksson has combined the Donnan and Vanselow theories and found:

$$\frac{[Na^+]^2(Ca^{2+})(C)}{(Na^+)^2[Ca^{2+}][Na^+ + Ca^{2+}]} = k \tag{7.11}$$

C = the exchange capacity of the colloid

7.11 EQUATIONS ACCORDING TO THE DIFFUSE DOUBLE-LAYER THEORY

Although considered by many to be the most realistic approach to describe ion exchange equilibria, the double-layer formula developed by Eriksson (1952) is far from simple. It has limited applications and works well for cation exchange equilibrium between Na and Ca but not between other ions. Therefore, for those who are in need of the equation, reference is made to Eriksson (1952) and Lagerwerff and Bolt (1959).

$$\frac{[Na^+]}{[Ca^{2+}]} =$$

$$\frac{[(1/\beta)/arc\ sinh](\beta)^{1/2}\Gamma/[(Na^+)/(Ca^{2+})^{1/2} + 4VC(Ca^{2+})^{1/2}]}{\Gamma - [(Na^+)/(Ca^{2+})^{1/2}/(\beta)^{1/2}]\ arc\ sinh(\beta)^{1/2}\Gamma/[Na^+)/(Ca^{2+})^{1/2}] + [4VC(Ca^{2+})^{1/2}]} R$$

$$R = \frac{(Na^+)}{(Ca^{2+})^{1/2}}$$

$$\beta = 8/100\ RT$$

$$VC = 1$$

Γ = charge density (mEq/cm^2) and arc sinh = $sinh^{-1}$

The diffuse double-layer equation can be reduced to a Gapon equation.

7.12 SCHOFIELD'S RATIO LAW

From the preceding discussion, it is apparent that although many equations are available to describe cation exchange phenomena, almost all of them have one thing in common. They all stated that the ratios of the products of adsorbed cations and cations free in solution are constant. This can be illustrated by the mass action and the Gapon equations. Both use the formula

$$\frac{[Na^+](\sqrt{Ca^{2+}})}{[\sqrt{Ca^{2+}}](Na^+)} = k$$

in which [ ] denotes adsorbed ions, and ( ) denotes ions in solution. Rearranging the equation gives

$$[Na^+](\sqrt{Ca^{2+}}) = k[\sqrt{Ca^{2+}}](Na^+)$$

or

$$\frac{(Na^+)}{(\sqrt{Ca^{2+}})} = \frac{1}{k}\frac{[Na^+]}{[\sqrt{Ca^{2+}}]} \tag{7.12}$$

The latter means that in equilibrium condition, the ratio of cations in solution depends on the ratio of cations adsorbed on the colloid surface. If the amount of cations adsorbed does not change significantly, or remains constant, the ratio of cations in solution $(Na^+)/(\sqrt{Ca^{2+}})$ is also constant. This is called the *ratio law* by Schofield (1947). The ratio law is used to predict the soil solution concentration as affected by fertilizers, lime application, or dilution by irrigation or rain. When upon dilution the Na^+ concentration decreases five times, the value of $(\sqrt{Ca^{2+}})$ must also decrease five times to obey the ratio law. This means that Ca^{2+} concentration then decreases 5^2 times. Hence, dilution favors a decrease in Ca content. If Na^+ increases five times, to maintain a constant ratio $(\sqrt{Ca^{2+}})$ must also increase five times. Therefore, Ca^{2+} concentration increases 5^2 times.

7.13 FIXATION OF CATIONS

Under certain conditions, the adsorbed cations are held so strongly by clays that they cannot be recovered by exchange reactions. These cations are called *fixed cations*. Although fixation can occur with almost any cation, the most important fixation reaction is with K^+ and NH_4^+ ions. Fixation of K^+ and NH_4^+ occurs by a similar mechanism. Among the several reasons reported for fixation, the most important is the entrapment of the ions in the intermicellar regions of the clays. Expanding lattice clays have octahedral holes of 1.40 Å in their intermicellar surfaces. When K^+ or NH_4^+ penetrates the intermicellar space, they will fit snugly into the holes. Upon closure of the space, the K^+ or NH_4^+ ions are trapped between the clay layers. They become, then, relatively nonexchangeable and are called *fixed* (Bolt et al., 1976; Rich, 1968; Van der Marel, 1959).

Many soil minerals have been reported to contribute to K^+ and NH_4^+ fixation (e.g., micas, illites, montmorillonites, and vermiculites). Van der Marel (1959) stated that permutites, zeolites, feldspars, and glauconite had the capacity to fix K. Rich (1968) was of the opinion that especially those minerals with strong interlayer charges and wedge zones exhibited sites for high potassium selectivity, in other words, exhibited high K fixation.

Although the quantity of K retained by fixation can assume high proportions, K fixation is currently considered more a disturbing than a harmful reaction. Depending on the conditions, significant amounts of the fixed K can be released and made available to plant growth. The presence of humic and fulvic acids in soils may accelerate this release (Tan, 1978b). Tisdale and Nelson (1975) are of the opinion that K fixation is a process of conservation in nature. Fixation is of special importance in sandy soils, where K is more likely to be lost rapidly by leaching. Continued application of K^+ or NH_4^+ fertilizers will decrease K fixation. The addition of K will fill the vacant positions in the clay lattice, in this way, satisfying the fixation capacity of soils.

7.14 BASE SATURATION

The *base saturation* is a property closely related to CEC. It is defined as

% Base Saturation

$$= \frac{\sum \text{Exchangeable Bases (in mEq/100 g)}}{\text{CEC}} \times 100\%$$

Example: Assume that in the analysis of soils the following concentration of bases and CEC value were obtained:

Exchangeable bases (mEq/100 g soil)	
Ca	10
Mg	5
K	10
Na	5
$\sum$ exchangeable bases = 30	

The CEC of the soil, determined separately, is 50 mEq/100 g. Therefore:

$$\% \text{ Base Saturation } = \frac{30}{50} \times 100 = 60\%$$

A positive correlation exists between percentage base saturation and soil pH. Generally, we can see that the base saturation is high if soil pH is high. Consequently, arid region soils are usually higher in base saturation than soils in humid regions. Low base saturation means the presence of a lot of H^+ ions.

The base saturation is frequently considered to be an indication of soil fertility. The ease with which adsorbed cations are released to plants depends on the degree of base saturation. A soil is considered very fertile if the percentage base saturation is $\geqslant 80\%$, medium fertile if percentage base saturation is 80–50%, and nonfertile if percentage base saturation is $\leqslant 50\%$. A soil with a percentage base saturation of 80% will release the exchangeable bases more easily than the same soil with a percentage base saturation of 50%. Liming is the common means by which the percentage base saturation of soils is increased.

8

Anion Exchange

8.1 POSITIVE CHARGES

In a preceding section, it was indicated that, under certain conditions, the soil colloids may also carry positive charges. This is especially true for the Fe- and Al-oxide minerals and amorphous soil colloids. Positive charges can also occur at the edges of clay minerals. This kind of charge is usually significant at pH values below the isoelectric point or zero point of charge. The broken-edged surface of an octahedral sheet, therefore, has a positively charged double layer at low pH. The double layer becomes increasingly more positive with decreasing pH.

8.2 ADSORPTION OF ANIONS BY SOIL COLLOIDS

Two types of adsorption of anions by soil colloids are recognized: negative and positive adsorption.

Negative Adsorption

Negative adsorption of anions occurs at a colloidal surface possessing a negative charge. Because of the latter, cations are attracted and concentrated at the colloid surface. On the other hand, anions are expelled from the double layer formed on the negatively charged surface. This exclusion of anions is called *negative adsorption*. Therefore, the bulk solution contains more anions than the solution in the interface. The amount excluded is reported to be a small part of the CEC. Bolt (1976) shows that under conditions prevailing in soil, the negative adsorption of anions is approximately 1–5% of the CEC. In contrast, negative adsorption can amount to 15% in saline soils.

Positive Adsorption

Positive adsorption of anions is the adsorption and concentration of anions on the positively charged surfaces or edges of soil colloids. Here, negative adsorption of cations (i.e., repulsion of cations by the positive charge) occurs.

The anion exchange capacity (AEC) of soils is usually smaller than the CEC. It is dependent on changes in electrolyte levels and on soil pH. It is also limited to special types of clays.

As with cations, lyotropic series of anions are also available. Bolt (1976) reported a decreasing order of preferential adsorption among the following anions:

$$SiO_4^{4-} > PO_4^{3-} \gg SO_4^{2-} > NO_3^- \approx Cl^-$$

This lyotropic series indicates that SiO_4^{4-} and PO_4^{3-} ions are strongly adsorbed; SO_4^{2-} and NO_3^- ions are adsorbed in considerably lower concentrations or are often not adsorbed. At a neutral soil reaction, or pH > 6, positive adsorption of SO_4^{2-} and Cl^- is very small. These two anions are generally subject to negative adsorption, in contrast with phosphate ions that are adsorbed more by positively charged surfaces or edges of clay minerals:

$$Al–OH + H_2PO_4^- \rightleftharpoons Al–H_2PO_4 + OH^-$$
(Clay)

This reaction is prevalent in acid soils. It results in a strong bond between the phosphate ion and the octahedral Al. Frequently, only part of the phosphate can be recovered by desorption analysis.

8.3 PHOSPHATE FIXATION AND RETENTION

Phosphate anions can be attracted to soil constituents with such a bond that they become insoluble and difficultly available to plants. This process is called *phosphate fixation* or *phosphate retention*. Many authors use the terms *fixation* and *retention* interchangeably. However, Tisdale and Nelson (1975) are of the opinion that *retention* refers to that part of adsorbed phosphorus that can be extracted with dilute acids. This fraction is relatively available to plants. The term *fixed*, on the other hand, is reserved for the portion of soil phosphorus that is not extractable by dilute acids. This portion of phosphorus is not readily available to plants. Under certain conditions, the distinction between phosphorus fixation and retention is rather obscure.

Phosphate Retention

Acid soils usually contain significant amounts of soluble and exchangeable Al^{3+}, Fe^{3+}, and Mn^{2+} ions. Phosphate, when present, may be adsorbed to the colloid surface with these ions serving as a bridge. This phenomenon is sometimes called *coadsorption*. The phosphate retained in this way is still available to plants. Such a reaction can also take place with Ca-saturated clays. Evidence has been shown that Ca clay adsorbs large amounts of phosphate. The Ca^{2+} ions form the linkage between the clay and phosphate ions:

$$\boxed{Clay} -Ca-H_2PO_4$$

According to the mass action or Gapon equation, the ionic concentration at the surface of the clay is dependent on that of the bulk solution:

$$[\sqrt{Ca^{2+}}][H_2PO_4^-] = (\sqrt{Ca^{2+}})(H_2PO_4^-)$$

in which [ ] denotes adsorbed species, and ( ) again equals free species. The phosphate ions can also enter into a chemical reaction with the foregoing free metal ions:

$$Al^{3+} + 3H_2PO_4^- \rightarrow Al(H_2PO_4)_3 \downarrow$$

The product formed is difficultly soluble in water and precipitates from solution. With the passage of time the Al phosphate precipitates become less soluble and less available to the plant.

The lower the soil pH, the greater the concentration of soluble Al, Fe, and Mn; consequently, the larger the amount of phosphorus retained in this way.

PHOSPHATE FIXATION IN ACIDIC SOILS

Fixation renders phosphate insoluble in water and relatively non-available to plants. The fixation reaction can occur between phosphate and Fe or Al hydrous oxides or between phosphate and silicate minerals.

Many acidic soils contain high amounts of free Fe and Al and of Fe and Al hydrous oxide clays, especially the highly weathered Ultisols in the United States and the Oxisols in the tropics. The free Fe, Al, and the sesquioxide clays react rapidly with phosphate, forming a series of difficultly soluble hydroxyphosphates:

Monodentate complex
(labile)

Bidentate complex
(irreversible)

The amount of phosphorus fixed by this reaction usually exceeds that fixed by phosphate retention. Such a reaction is not limited to Al and Fe hydrous oxide clays, but Mn hydrous oxide clays and amorphous clays also have considerable phosphate fixing capacities. In contrast with phosphate retention that occurs mainly in acidic soil condition, phosphate fixation by hydrous oxide clays occurs over a relatively wider pH range.

The products formed by both retention and fixation reactions are frequently not pure Al or pure Fe phosphates. The ultimate end product of the reaction between aluminum hydroxides and phosphates is called *variscite* ($AlPO_4 \cdot 2H_2O$) and that of Fe and phosphate is known as *strengite* ($FePO_4 \cdot 2H_2O$). A series of intergrades between variscite and strengite is usually present in soils and is called the variscite–strengite isomorphous series (Lindsay et al., 1959).

Another type of phosphate fixation is the reaction between phosphate and silicate clays. Especially soil clays exhibiting exposed OH groups, such as the kaolinitic groups, have a strong affinity for phosphate ions. Generally, clays with low SiO_2/R_2O_3 (sesquioxide) ratios have a higher phosphate-fixing capacity than clays with high SiO_2/R_2O_3 (sesquioxide) ratios.

The phosphate ions react rapidly with octahedral Al by replacing the OH groups located on the surface plane of the mineral. This type of reaction is also prevalent in acidic conditions.

Phosphate Fixation in Alkaline Soils

Fixation of phosphate is not limited to acidic conditions only, but also occurs readily in alkaline soil reactions. Many alkaline soils contain high amounts of soluble and exchangeable Ca^{2+} and, frequently, $CaCO_3$. Phosphate is reported to react with both the ionic and the carbonate form of Ca. The latter can be illustrated as follows:

$$3Ca^{2+} + 2PO_4^{3-} \rightarrow Ca_3(PO_4)_2 \downarrow$$
$$\text{(insoluble)}$$
$$3CaCO_3 + 2PO_4^{3-} \rightarrow Ca_3(PO_4)_2 + 3CO_2 \uparrow$$
$$\text{(insoluble)}$$

Other forms of insoluble Ca phosphate can also be formed by this type of reaction between calcium and phosphate (e.g., hydroxy-, oxy- and Ca-fluoroapatite [$Ca_5(PO_4)_3F$]).

Such type of fixation is a serious problem in arid region soils of the western part of the United States. However, it can also become significant in the humid region of the eastern seaboard of the United States when the soils receive high applications of lime. In these soils or under such conditions application of P fertilizers generally gives low plant growth response. Phosphate fixation cannot be avoided entirely, but it may be reduced by addition of competing ions for fixing sites. Organic anions from stable manure and silicates are reported to be very useful in reducing P fixation (Bolt, 1976).

8.4 SOIL REACTION AND AVAILABILITY OF INORGANIC PHOSPHATES

As just discussed, insoluble Al and Fe hydroxyphosphates are stable in acidic conditions, whereas Ca phosphates are prevalent in alkaline conditions. At pH 3–4, solubility of the Al and Fe hydroxyphosphates is considered very low. However, with increasing pH levels, solubility of these phosphate compounds increases and reaches a maximum at approximately pH 6.5. Above this pH level, the Al- and Fe-hydroxy compounds decrease again in solubility (Figure 8.1). On the extreme alkaline range, pH 8.0, Ca phosphate is in an insoluble form. By decreasing the pH, this compound also becomes slightly soluble, and maximum solubility is reached at pH 6.5.

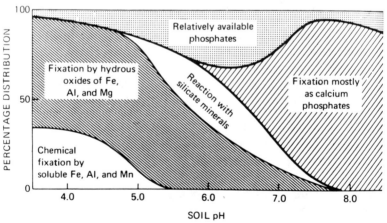

Figure 8.1 Phosphate availability and fixation as related to soil pH. (From Brady, 1974.)

Therefore, it appears that at pH 6.5, the soil may contain the maximum amounts of phosphate that can be solubilized from all the insoluble inorganic phosphate forms present in soils. The forms of phosphate ions present also depend on soil pH. It is known that in acidic conditions the $H_2PO_4^-$ ions prevail, in alkaline conditions the HPO_4^{2-} ions are dominant, and at pH 6.5, $H_2PO_4^-$, HPO_4^{2-}, and PO_4^{3-} can exist in combination in the soil solution.

8.5 PREDICTION OF PHOSPHATE ION CONCENTRATION ACCORDING TO DOUBLE-LAYER THEORIES

From the foregoing, it is apparent that the concentration of phosphate ions in the soil solution depends upon several factors. It is usually affected by soil pH, by some member of the variscite–strengite isomorphous series, and by a member of the Ca–hydroxyphosphate series. According to Schofield's ratio law, the following relations are valid:

$$(H^+) = \frac{K_w}{(OH^-)} \tag{8.1}$$

$$\frac{[\sqrt{Ca^{2+}}]}{[H^+]} = \frac{(\sqrt{Ca^{2+}})}{(H^+)} = k_1 \tag{8.2}$$

$$\frac{(Al^{3+})^{1/3}}{(Ca^{2+})^{1/2}} = k_2 \tag{8.3}$$

Lindsay et al. (1959) assume that if variscite, $Al(OH)_2H_2PO_4$, will dissociate completely into its component ions, we can write the solubility product (K_{sp}) of variscite as follows:

$$K_{sp} = (Al^{3+})(OH^-)^2(H_2PO_4^-)$$

By using Eqs. (8.1)–(8.3) for substitution into the equation for K_{sp}, Lindsay et al. (1959) stated,

$$(\sqrt{Ca^{2+}})(H_2PO_4^-) = \text{constant}$$
$$(H^+)(H_2PO_4^-) = \text{constant}$$
$$(Al^{3+})^{1/3}(H_2PO_4^-) = \text{constant}$$

8.6 THE PHOSPHATE POTENTIAL

Three types of phosphate potentials can be distinguished, chemical potential of phosphate, electrochemical potential of phosphate, and Schofield's phosphate potential.

Chemical Potential of Phosphate

As discussed earlier, each chemical species carries a certain amount of energy. The unit amount of free energy per unit amount of chemical species was defined earlier as the *chemical potential*. If this concept is applied with a phosphate ion, $H_2PO_4^-$, the chemical potential of this phosphate ion can be formulated as

$$\mu = \mu^0 + RT \ln a \tag{8.4}$$

where μ = chemical potential of $H_2PO_4^-$, μ^0 = standard chemical potential of $H_2PO_4^-$, R = gas constant, T = absolute temperature in Kelvin, and a = activity of $H_2PO_4^-$. Conversion in log form gives

$$\mu = \mu^0 + 1.364 \log a$$

According to this concept, movement of phosphate ions from one to another location in soils is always in the direction of lower chemical potentials. Therefore, phosphate ions are taken up by plant roots if the chemical potential of phosphate ions in soils is larger than the chemical potential of phosphate ions in plants:

$$\mu_{P(plants)} < \mu_{P(soil)}$$

Electrochemical Potential of Phosphate

By definition, the *electrochemical potential* is the sum of the chemical potential and electrical potential of phosphate, which can be formulated as follows:

$$\phi = \mu + zF\psi \tag{8.5}$$

in which ϕ = electrochemical potential of a phosphate ion, μ = chemical potential of the phosphate ion, z = valence, F = Faraday constant, and ψ = electrical potential.

When phosphate uptake by plant roots reaches an equilibrium, movement of phosphate ions will stop. At equilibrium, the system obeys the *Donnan equilibrium law*. The electrochemical potentials of phosphate ions within the plant cells (ϕ_i) must then equal the electrochemical potentials of phosphate ions in the soil solution (ϕ_o):

$$\phi_i = \phi_o$$
$$\mu_i + (zF\psi)_i = \mu_o + (zF\psi)_o$$
$$(RT \ln a)_i + (zF\psi)_i = (RT \ln a)_o + (zF\psi)_o$$
$$(zF\psi)_o - (zF\psi)_i = (RT \ln a)_i - (RT \ln a)_o$$
$$\psi_o - \psi_i = \frac{RT}{zF} \ln \frac{a_i}{a_o}$$

which has previously been called the *membrane or Donnan potential*, a_i = phosphate ion activity inside the plant cells, and a_o = phosphate ion activity in the soil solution (outside plant cells).

Changing from the natural to the common logarithm (ln = 2.303 log) and substituting R and F for the gas and Faraday constant, respectively, the Donnan potential assumes at $T = 298$ K (25°C) the following formula:

$$E = \psi_o - \psi_i = 0.059/z \log a_i/a_o \tag{8.6}$$

Schofield's Phosphate Potential

The term *phosphate potential* is introduced by Schofield (1955) to be used as an index for availability of soil phosphorus. Similarly, as with the definition of soil water potential, the *phosphate potential* is defined as the amount of work that must be conducted to move reversibly and isothermally an infinitesimally small amount of a phosphate ion from a pool of phosphates at a specified location at atmospheric pressure to the point under consideration.

Schofield (1955) was of the opinion that a labile pool of phosphates existed in the soil from which the plant roots could draw their P needs. By measuring the pH and the total P concentration in 0.01 M $CaCl_2$ solution, the solubility of phosphate can be expressed as:

$$(Ca^{2+})^{1/2}(H_2PO_4^-)$$

By taking $-\log$, this product changes into

$$-\tfrac{1}{2} \log(Ca^{2+}) + -\log(H_2PO_4^-)$$

or

$$\frac{1}{2}\,p\text{Ca} + p\text{H}_2\text{PO}_4 \tag{8.7}$$

in which $p = -\log$. Equation (8.7) is called the *phosphate potential*. In analogy to the water potential, a low phosphate potential suggests high availability, whereas a high phosphate potential refers to lower availability of P to plants. Because several technical difficulties, the use of phosphate potentials as an index of availability of P has not been tested. However, since Eq. (8.7) is, in fact, a solubility product, it is expected to be useful in estimating solubility of phosphates in soil solutions. Since availability of P to plants is related to solubility of P, the phosphate potential can be used to make indirect predictions about the phosphate availability to plants.

9

Soil Reaction

9.1 DEFINITION AND IMPORTANCE

The *soil reaction* is a term used to indicate the acid–base reactions in soils (Tisdale and Nelson, 1975; Brady, 1974). Several soil processes are affected by the soil reaction. Many soil chemical and biochemical reactions can occur only at specific soil reactions. The rate of decomposition of soil minerals and organic matter is influenced by the soil reaction. Formation of clay minerals depends on the soil reaction. Plant growth is also affected, either directly or indirectly, by the acid–base reactions in soils. The latter may influence plant growth indirectly through its effect on solubility and availability of plant nutrients. The changing phosphate concentration with soil pH, as discussed earlier, is a good example. Directly, H^+ ions are reported to have a toxic effect on plants when present in high concentration.

The colloidal particles can also behave as an acid or a base. Hydrogen or Al-saturated clays usually behave as an acid and may react with bases.

9.2 ACID–BASE CHEMISTRY

To understand and manipulate the soil acid–base system, it is important to first define what an acid or a base is. Three major concepts of acids and bases are available: Arrhenius, Brønsted, and Lewis concepts.

Arrhenius Concept

The concept of Arrhenius developed between 1880 and 1890 stated that an *acid* is a compound that contains hydrogen. In aqueous solution the acid yields hydrogen ions (H^+). A *base* is a compound that produces hydroxyl ions (OH^-) in solutions. The Arrhenius concept is essentially valid only for the definition of acids. Almost all acids contain hydrogen. However, Arrhenius' definition of a base limits the base to only compounds with hydroxyl ions. It is currently known that ammonia (NH_3) and many other organic substances exhibit characteristics of base compounds.

Brønsted–Lowry Concept

Brønsted and Lowry defined separately, in 1923, that an *acid* is a compound capable of donating a proton (proton donor). On the other hand, any compound capable of accepting a proton is considered a *base* (proton acceptor). If we take as an example HCl, which is usually considered an acid, then according to the Brønsted–Lowry theory, the compound can donate a proton. Therefore, HCl is an acid. However, after dissociation the remaining Cl^- ion is then a base, because it can accept a proton. This type of acid–base pair is called a *conjugate pair* with the Cl^- being the conjugate base of the acid HCl.

If HCl is dissolved in water, the following reaction occurs:

$$HCl + H_2O \rightarrow H_3O^+ + Cl^-$$

H_3O^+ is called a *hydronium ion*. Since hydronium is formed by adsorption of a proton by the water molecule, water is a proton acceptor. Consequently water can be considered a base. The reaction to the left ($H_2O + Cl^-$) will not occur since H_2O is a stronger proton acceptor than proton donor.

Lewis Concept

According to this theory, also developed in 1923, an acid is a compound that can accept an electron pair. A base is a compound that can donate an electron pair. The following serves as an example:

$$H^+ + :\ddot{O}: H^- \rightarrow H :\ddot{O}: H$$

In this equation, the H^+ ion accepts an electron pair. The proton is then the acid and the bonding is called a *covalent bond*. The hydroxide ion donates the electron pair and is considered the base. Another example is as follows:

$$H^+ + :\overset{H}{\underset{H}{\ddot{N}}}: H \rightarrow H :\overset{H}{\underset{H}{\ddot{N}}}: H$$

Again, the H^+ ion accepts an electron pair and is the acid. NH_3 is then the base, since it donates the electron pair.

In studying these three theories, perhaps it can be noted that in the soil solution at one time both the Brønsted–Lowry and the Lewis theories can be applied. However, at another time, the Arrhenius and Brønsted–Lowry theories are more suitable to describe the condition. Reactions with clay minerals perhaps follow the Brønsted–Lowry concept more, whereas complex reactions involving organic matter apply only to the Lewis theory. If clay can adsorb and dissociate protons as illustrated by the reaction:

$$Clay\text{-}H_x \rightleftharpoons H^+$$

where H_x = number of adsorbed H^+, we can assume that soil clays can

1. Accept H^+. According to the Brønsted–Lowry concept clay is then a base.
2. Dissociate H^+. Applying again the Brønsted–Lowry theory, clay is an acid.

Hence, clay then behaves as an amphoteric compound.

9.3 FORMULATION OF SOIL ACIDITY AND ALKALINITY

Soil pH

The theories of Arrhenius and Brønsted in combination must be applied to characterize acid and alkaline conditions in soils. In acidic soils, more H^+ than OH^- ions are present. On the other hand, a basic soil has in its soil solution more OH^- than H^+ ions. To characterize these conditions the term *soil pH* is used. The term pH was introduced by Sörensen in 1909 and is defined as

$$pH = \log \frac{1}{A_{H^+}} = -\log A_{H^+} \tag{9.1}$$

where A_{H^+} = activity of H^+ ions.

However, frequently it is more convenient to use H^+ ion concentration rather than activity, and Eq. (9.1) becomes

$$pH = \log \frac{1}{(H^+)} = -\log(H^+)$$

Application of this pH concept in the dissociation of pure water gives the following relationship. As discussed earlier the ion product of water is

$$K_w = C_{H^+} \times C_{OH^-} = 10^{-14} \text{ at } 25°C$$

By taking the $-\log$, this equation changes into

$$-\log C_{H^+} - \log C_{OH^-} = -\log 10^{-14}$$

or

$$-\log(H^+) - \log(OH^-) = 14$$

or

$$pH + pOH = 14 \tag{9.2}$$

in which p $= -\log$; therefore pOH $= -\log$ of the hydroxyl ion concentration. It (pOH) is calculated in a similar manner as pH.

Equation (9.2) states that pH + pOH = constant and conversion of one into the other is a simple matter. Therefore, in describing soil acidity or alkalinity it is not necessary to determine both pH and pOH. If pH is known, pOH can be calculated using Eq. (9.2).

Acidity Constant

According to the Brønsted–Lowry concept the following acid–base relationship is valid:

$$\text{Acid} \rightleftharpoons \text{H}^+ + \text{base}$$

Since this is a dissociation reaction, at equilibrium we therefore have

$$K_A = \frac{(\text{H}^+)(\text{base})}{(\text{acid})}$$

in which K_A is called the *acidity constant*. Application of the pH concept in this equation gives

$$pK_A = \frac{\text{pH}[-\log(\text{base})]}{-\log(\text{acid})}$$

or

$$\text{pH} = pK_A + \log \frac{(\text{base})}{(\text{acid})}$$

The latter means that for a given ratio of concentration (or activities) of an acid and its conjugated base, the pH has a fixed value (Novozamsky et al., 1976).

Acid Strength and Ion Pairs

The strength of an acid HB depends on the reaction of the acid with the solvent:

$$\underset{\text{acid}}{\text{HB}} + \underset{\text{solvent}}{\text{HS}} \rightarrow \underset{\text{ion pair}}{\text{H}_2\text{S}^+\text{B}^-}$$

This reaction results in ionization of the acid and conjugated base.

In most electrolytes, the resulting component ions do not completely dissociate from each other. The cations and anions are strongly attracted to each other, and a large part behaves as if it is not ionized. These ions present in association are called *ion pairs* (Davies, 1962). The degree of ionization is dependent on the relative alkaline strengths of the conjugated base and the solvent.

The ion pair will dissociate into its component ions:

$$H_2S^+B^- \rightleftharpoons H_2S^+ + B^-$$
(dissociation)

The degree of dissociation depends on the dielectric constant of the solvent. (The dielectric constant is the ratio of the capacity of an electric condenser in a vacuum and in the solvent, $\epsilon = C_v/C_s$.)

By combining the two foregoing reactions the dissociation constant (K_A) of HB can be written as

$$K_A = \frac{(H_2S^+)(B^-)}{(HB) + (H_2S^+B^-)}$$

The value for K_A is used as a quantitative measurement of the strength of the acid.

9.4   CONCEPTS OF SOIL ACIDITY

Soil pH Range

On the basis of their relative degree of acidity, the soils are divided into several acidity or alkalinity classes, as shown in Figure 9.1. Acidic soils are usually common in humid regions. In these soils, the concentration of H^+ ions exceeds that of OH^- ions. These soils may contain large amounts of soluble Al, Fe, and Mn. Alkaline soils occur mostly in semiarid to arid regions. Here the OH^- ions are dominant over the H^+ ions. Because of the alkaline reaction, the soils contain low amounts of soluble Al, Fe, and Mn.

Active Versus Potential Soil Acidity

A number of compounds contribute to the development of acidic or basic soil reactions. Inorganic and organic acids, produced by the decomposition of soil organic matter, are common soil constituents

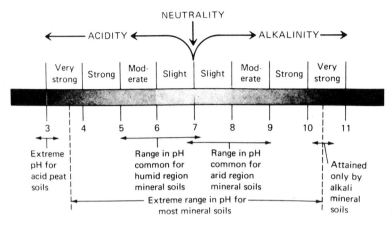

Figure 9.1 Soil pH ranges or soil reaction classes. (From Brady, 1974.)

that may affect soil acidity. Respiration of plant roots yield CO_2 that will produce H_2CO_3 in water. Water is another source for a small amount of H^+ ions. A large portion of the H^+ ions present in soils will be absorbed by the clay complex as exchangeable H^+ ions:

$$\boxed{Clay}\begin{array}{l} H \\ H \\ H \end{array} \rightleftharpoons H^+$$

Potential   Active
   or
reserve

These exchangeable H^+ ions dissociate into free H^+ ions. The degree of ionization and dissociation into the soil solution determines the nature of soil acidity. The exchangeable H^+ ions are the reason for the development of potential or reserve soil acidity. The magnitude of the latter can be determined by titration of the soil. The free H^+ ions create the active acidity. Active acidity is measured and expressed as soil pH. This is the type of acidity upon which plant growth reacts.

Nonselective and Preferential Adsorption of Hydrogen Ions by Soils

Bolt (1976) distinguishes two types of adsorption of H^+ ions by soil colloids (i.e., nonselective and preferential adsorption). Soil colloids, such as clay, adsorb H^+ ions by the nonselective process. They accumulate on the charged surface as a swarm of counterions. The relative proportion can be estimated with the Gapon equation.

On the other hand, organic colloids exhibit preferential adsorption of H^+ ions. The organic compounds contain acidic groups that are highly selective for association with protons. The adsorbed H^+ ions are, then, considered part of the group, or the surface. Bolt (1976) considered them more difficulty exchangeable against other cations.

9.5 THE ROLE OF ALUMINUM IN SOIL ACIDITY

Most clay particles interact with H^+ ions. Evidence is available that a hydrogen-saturated clay undergoes a spontaneous decomposition. The hydrogen ion penetrates the octahedral layer and replaces the Al atoms. The Al released is then adsorbed by the clay complex and a H–Al–clay complex is formed rapidly. The Al^{3+} ions may hydrolyze and produce H^+ ions:

$$\boxed{\text{clay}}\ Al\ +\ 3H_2O \rightarrow Al(OH)_3 \downarrow\ +\ H \overset{\text{H}}{\underset{\text{H}}{\boxed{\text{clay}}}} \rightleftharpoons H^+$$

This reaction contributes toward increasing the H^+ ion concentration in soils. The magnitude of the change in pH is related to the Al concentration expressed in terms of the *aluminum potential*, pAl, which can be explained as follows. Hydrolysis of Al ions can be illustrated by the following step-wise reactions:

$$Al^{3+}\ +\ H_2O \rightarrow Al(OH)^{2+}\ +\ H^+$$
$$Al(OH)^{2+}\ +\ H_2O \rightarrow Al(OH)_2^+\ +\ H^+$$
$$\underline{Al(OH)_2^+\ +\ H_2O \rightarrow Al(OH)_3\ +\ H^+}$$
$$Al^{3+}\ +\ 3H_2O \rightarrow Al(OH)_3\ +\ 3H^+$$

Applying the mass action to the total or final reaction gives

$$K_h = \frac{(Al(OH)_3)(H^+)^3}{(Al^{3+})(H_2O)^3}$$

where K_h = hydrolysis constant. By considering $Al(OH)_3$ and H_2O pure compounds at standard states, their activity = unity. Hence, the equation can be converted into:

$$K_h = \frac{(H^+)^3}{(Al^{3+})}$$

By taking $-\log$, the equation becomes:

$$-\log K_h = -\log(H^+)^3 - -\log(Al^{3+})$$
$$pK_h = 3pH - pAl^{3+}$$
$$3pH = pK_h + pAl^{3+}$$

Division by 3 gives:

$$pH = \tfrac{1}{3}pK_h + \tfrac{1}{3}pAl^{3+}$$

This equation, often called the *aluminum potential*, shows the relation between pH and Al activity in soils.

Polymerization of the Al-monomers, $Al(OH)^{2+}$, yields dimeric Al-hydroxides:

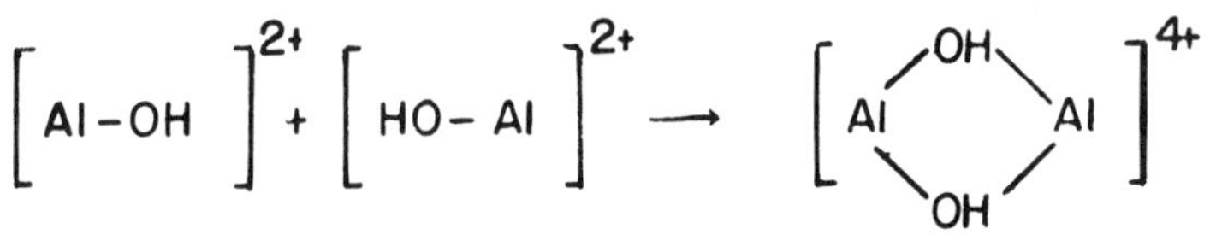

Olation reaction

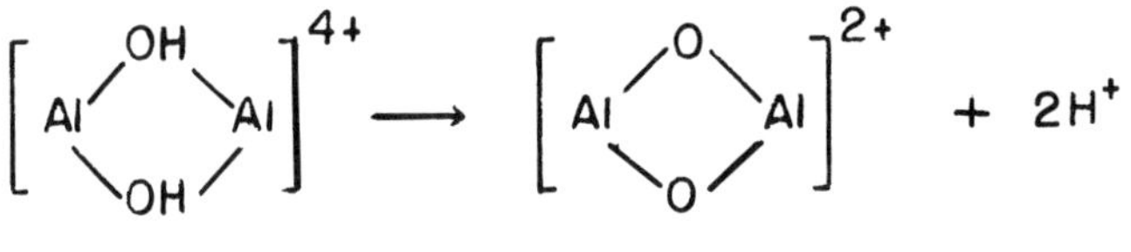

Oxolation reaction

As illustrated, oxolation reaction of Al-hydroxide polymers also increases the H^+ ion concentration of the soil solution.

9.6 THE ROLE OF FERTILIZERS, PYRITE, AND ACID RAIN IN SOIL ACIDITY

In addition to the aforementioned processes, both potential and active acidity also can be increased by the following processes:

Large Applications of Ammonium Fertilizers

In a normal healthy soil, ammonium ions will be attacked by microorganisms and oxidized into nitrate ions according to the reactions:

$$2NH_4^+ + 3O_2 \rightarrow 2NO_2^- + 2H_2O + 4H^+ + \text{energy}$$

$$2NO_2^- + O_2 \rightarrow 2NO_3^- + \text{energy}$$

As indicated in an earlier chapter, the oxidation of ammonium into nitrate, called *nitrification*, occurs into two steps. In the first step, the bacterial conversion of ammonium into nitrite by *Nitrosomonas* sp., four protons are produced, which may increased soil acidity. In the second and final step, nitrite is bacterially oxidized into nitrate by *Nitrobacter* sp.

Application of Monocalcium Phosphate Fertilizers

Hydrolysis of this phosphate fertilizer will yield orthophosphoric acid:

$$Ca(H_2PO_4)_2 \rightarrow CaHPO_4 + H_3PO_4$$

Being an acid, *o*-phosphoric acid will dissociate its proton in the soil solution, and decreases soil pH.

Oxidation of Pyrite (FeS_2), a Sulfur Mineral

$$2FeS_2 + 7H_2O + 7\tfrac{1}{2}O_2 \rightarrow 4SO_4^{2-} + 8H^+ + 2Fe(OH)_3$$

This reaction produces eight protons and contributes toward decreasing soil acidity. The process is of special importance in soils derived from mine spoils and in drained areas affected by the tide. Coal contains large amounts of sulfur in the form of pyrite and organic sulfur. The use of coal as an energy source will oxidize the pyrite minerals. Not only will this produce protons, but also sulfur dioxide gas, which is an important contributor to acid rain, as will be discussed in the next section.

Acid Precipitation

Acid precipitation, also known as *acid rain*, is caused by the conversion of nitrogen oxide and sulfur dioxide gases into strong acids. The process is actually a natural process and occurs frequently in regions with active volcanism. Carbon monoxide, nitrogen oxide, and sulfur oxide gases are released into the air by active volcanoes and contribute to acid rain and acid soils in the region. However, it is the enhanced and continuous production of unwanted gases, generated by the combustion of fuel from motor vehicles, electric power production, space heating and cooling, and disposal of refuse, that has become of global concern. In the United States alone, it was estimated, in 1965, that those processes have produced and released in the air 72 million tons of carbon monoxide, 26 million tons of sulfur dioxides, and 13 million tons of nitrogen oxides (Manahan, 1975). These gases are particularly harmful to the environment, plant growth, and human health. Carbon monoxide is very toxic for animals and human beings. It displaces oxygen from hemoglobin and reduces the capacity of blood to carry oxygen. The latter results in suffocation. Exposure to high levels of sulfur dioxide gas may cause *leaf necrosis* in plants. There is now evidence that nitrogen oxide gases, produced not only by the industry, but also by supersonic and space transport, may be harmful to the ozone layer of the stratosphere. From the standpoint of acid rain, carbon monoxide is of less significance than sulfur dioxides and nitrogen oxides.

Oxidation of sulfur dioxides and nitrogen oxides in the air, and subsequent dissolution of the oxidation products in raindrops, produce sulfuric acids, nitric acids, and their salts. Not only are these acids very corrosive, but they also result in a decrease in pH of the rainwater. The salts of these acids have been implicated in the formation of turbid haze or fog covering the industrial towns in the

Midwest and California. The processes of formation of acids can be illustrated by the following reactions:

$$2SO_2 + O_2 \rightarrow 2SO_3$$

$$SO_3 + H_2O \rightarrow H_2SO_4$$
$$\text{(sulfuric acid)}$$

$$2NO + O_2 \rightarrow 2NO_2$$

$$2NO_2 + H_2O \rightarrow HNO_3 + HNO_2$$
$$\text{(nitric and nitrous acids)}$$

Sulfuric, nitric, and nitrous acids are strong acids and will dissociate their protons in water droplets of fog or rain. Because of this, the pH of rainwater or water in fog may decrease to 2.0 or lower. Acid rain has become of increasing concern today in Europe and North America because of *die-back* of forest trees, a harmful process also known as *forest decline*. Reports have also indicated that the increased acidity of lake water in the Adirondack Mountains of the United States, because of acid rain, has contributed to *fish kill*, since most fish are sensitive to pH below 4.5. However, the effect of acid rain is of a less serious nature in soils than in lake water. Soils exhibit a cation exchange capacity (CEC) that provides them with a buffering complex to adsorb the excess protons from acid rain. However, a prolonged impact by acid rain can saturate the buffer capacity of soils, which eventually increases soil acidity. This is especially critical in sandy soils that are weakly buffered by nature. Two methods have been proposed to control the effect of acid rain on soil acidity (i.e., cleaning the emissions in the industry and motor vehicles, or liming the soils, or both). For example, the primary source of enhanced production of sulfur dioxide gas is coal. The sulfur in the coal is present as pyrite, FeS_2, and as organic sulfur. Upon burning the coal, the pyrite minerals are oxidized and sulfur dioxide is produced, as discussed in a previous section. Removing the S from the coal before use in the ovens, to keep sulfur dioxide emissions at relatively low level, is still a costly process.

9.7 BUFFERING CAPACITY OF SOILS

Chemically, a *buffer solution* is defined as one that resists a change in pH on addition of acid or alkali. Buffer solutions contain compounds that react with both acid or base so that the H^+ ion concentration in the solution remains constant.

In soils, the clay and humic fractions act as a buffer system. As discussed previously, the soil cation exchange complex creates the development of potential and active acidity. The potential acidity will maintain the equilibrium with the active acidity. If the active H^+ ion concentration is neutralized by the addition of lime, the potential acidity will release exchangeable H^+ ions into the soil solution to restore the equilibrium, and no change in soil reaction occurs until the reserve in H^+ is exhausted. The magnitude of the potential acidity usually far exceeds that of the active acidity. Brady (1974) reported that in sandy soils, reserve acidity (H^+ on the exchange complex) was 1000 times greater than active acidity. In clay soils high in organic matter, reserve acidity was even 50,000–100,000 times greater than active acidity. Therefore, buffering capacity is greater in clay soils than in sandy soils. The larger the buffer capacity, the larger the amounts of lime needed to raise the soil pH to the desired level.

The concept of buffering capacity of soils is not limited to the soil's resistance to changes in soil reaction. The soil can also act as a filter for dissolved and colloidal contaminants. It may act as a sieve, or during the passage through the top soil, the aerated condition may oxidize and mineralize, in particular, the organic compounds. The ions released by mineralization are adsorbed by the soil adsorption complex and prevented from reaching the groundwater.

9.8 ELECTROMETRIC MEASUREMENT OF SOIL pH

The determination of soil pH is now made with the glass electrode. It consists of a thin glass bulb containing dilute HCl, into which is inserted an Ag–AgCl wire, serving as the electrode with a fixed voltage. When the glass bulb is immersed in a solution, a potential difference develops between the solution in the bulb and the soil solution outside the bulb. This potential E is formulated by the *Nernst* equation:

$$E = \frac{RT}{nF} \log \frac{K}{M^{n+}}$$

Ag–AgCl

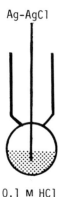

0.1 M HCl

where

R = gas constant
T = absolute temperature
n = valence
F = Faraday constant
K = constant
M = activity of ions to be measured

E is called the *half-cell potential* and cannot be measured alone. If the glass electrode is placed against a reference electrode (usually the calomel electrode), the potential difference between the two ($E - E_{cal}$) is measurable.

Before any pH measurement, the two electrodes have to be placed first in a solution of known pH (e.g., H^+ ion concentration = 1 g/L). This is called *standardizing the electrodes* and the pH meter. The overall potential of the total cell E_0 equals $E - E_{cal}$, and is as follows:

$$E_0 = \frac{RT}{F} \log \frac{K}{1} - E_{cal} \tag{9.3}$$

If the two electrodes are now placed in the solution with the unknown H^+ ion concentration, the potential E_c is

$$E_c = \frac{RT}{F} \log \frac{K}{(H^+)} - E_{cal} \tag{9.4}$$

Substracting Eq. (9.3) from Eq. (9.4) gives

$$E_c - E_0 = \frac{RT}{F}\left(\log\frac{K}{(H^+)} - \log K\right)$$

$$= \frac{RT}{F}\log\frac{1}{H^+}$$

or

$$E_c - E_0 = \frac{RT}{F}\,pH$$

For H^+ ions, $RT/F = 0.0591$ at 25°C; therefore

$$pH = \frac{E_c - E_0}{0.0591} \tag{9.5}$$

The soil pH can be measured in several ways. It can be measured in a water extract of soil, aqueous soil suspension, KCl–soil suspension, or CaCl–soil suspension.

Measurement of pH in a Water Extract

Soil and water are mixed in 1:1 or 1:2 (weight/volume) ratio, and stirred thoroughly for 15 min, after which the supernatant is collected by centrifugation for pH measurement. The pH value obtained usually increases with increased volume of water used. The increase in pH is caused by the dilution of the H^+ ion concentration in the solution. Therefore, the smallest soil/water ratio, that can be used without producing technical difficulties, is preferred in this method of pH measurement. Some workers prefer to use a saturated soil paste. However, in sandy soils the contact between the electrode and the soil solution may be decreased, and the chance of damaging the electrode is increased.

Measurement of pH in a Soil Suspension

Soil and water are also mixed in 1:1 or 1:2 ratio, stirred for 15 min, after which the pH is directly measured in the soil suspension. As with the previous method, the value obtained will also change with increased dilution by the use of wider soil/water ratios. The value obtained is considered to be closer to the true pH value in nature than the supernatant pH. It is often lower in value than the soil pH measured in its supernatant solution. This is caused by the so-called suspension effect, which will be discussed in a separate section.

Measurement of Soil pH in a Potassium Chloride Solution

For this method, the soil is mixed with a 1 M KCl solution in 1:1 or 1:2 ratio. The pH is measured directly in the soil suspension. This type of measurement will yield a more stable result than with the two other previously discussed methods. It is sometimes referred to as the *buffered pH*. In acid soils, the pH value obtained is usually lower than that measured with the other methods. The use of pH_{KCl} will provide better information concerning the chemical properties of the system (Moore and Loeppert, 1987). It is believed that the KCl pH reflects closely the CEC, and the cationic composition of the exchange complex.

Measurement of Soil pH in a Calcium Chloride Solution

The soil is mixed with a 0.1 M $CaCl_2$ solution using a 1:1 or 1:2 soil/solution ratio, and the pH is measured in the $CaCl_2$ suspension. This type of measurement is often used in conjunction with the determination of the *lime potential*, which will be discussed separately later.

9.9 SUSPENSION EFFECT IN SOIL pH MEASUREMENT

In the pH measurement, the reference and indicator electrodes are immersed in a heterogeneous soil suspension composed of dispersed solid particles in an aqueous solution. If the solid particles are allowed to settle, the pH can be measured in the supernatant liquid or in the sediment. Placement of the electrode pair in the supernatant usually gives a higher pH reading than placement of the electrodes in the sediment. This difference in soil pH reading is called the *suspension effect*.

Stirring the soil suspension before measurement will not solve the problem, since the latter procedure gives unstable readings.

According to Bolt et al. (1976), the reference electrode should be placed in the supernatant. The position of the glass electrode is considered immaterial at equilibrium condition.

9.10 LIME POTENTIAL

It is apparent from the foregoing discussions that measurements of soil pH can give highly variable results, since they are affected by several factors (e.g., suspension effect, soil/water ratios, and electrolyte levels). Schofield and Taylor (1955) proposed the use of 0.1 M $CaCl_2$ solutions for obtaining stable readings in pH measurements. However, instead of using single ion activity measurements, Schofield and Taylor (1955) suggested the use of ion activity ratios for determination of soil acidity and base saturation. If the soil exchange complex is saturated with both H^+ and Ca^{2+} ions, at equilibrium Schofield's ratio law says that

$$\frac{(H^+)}{(\sqrt{Ca^{2+}})} = \text{constant}$$

By taking $-\log$, this ratio changes to

$$-\log \frac{(H^+)}{\sqrt{Ca^{2+}})} = \text{constant}$$

$$-\log(H^+) - [-\log(\sqrt{Ca^{2+}})] = \text{constant}$$

or

$$pH - \tfrac{1}{2}pCa = \text{constant} \tag{9.6}$$

This equation is called the *lime potential*. It characterizes the composition of the exchange complex relative to its saturation by H^+ and Ca^{2+} ions. Many of the methods for determination of lime requirement take into consideration the lime potential.

9.11 THE NEED FOR ACIDIC SOIL REACTIONS

Usually, soil fertility is improved by liming acidic soils to pH 6–7. Most plants grow well in this pH range. At this soil reaction, available Ca, Mg, and P concentrations are adequate for plant growth. The levels of micronutrient contents in the soil solution are sufficient. Both fungal and bacterial activity are also present.

However, in certain cases it is desirable to maintain a strong to moderately acidic condition for plant growth. Some ornamental plants, (e.g., azaleas and rhododendrons) require acidic soil reactions for optimum growth. Pine trees also grow better in acidic soils. Other crops that are grown in acidic soils are blueberries, pineapples, and Irish potatoes. With potatoes, the acidic condition will reduce the development of potato scab, a disease caused by actinomycetes.

A number of compounds can be used to maintain or intensify soil acidity. Acid organic matter (e.g., pine needles) or chemical compounds such as S powder or $FeSO_4$ can be mixed with the soil to produce the acid reaction.

9.12 SOIL REACTIONS IN SALINE AND SODIC SOILS

Salinization

Saline and sodic soils, today called *Aridisols*, are soils of arid regions where the average precipitation is less than 500 mm (20 in.) annually. The amount of H_2O coming from the precipitation is insufficient to neutralize the amount of H_2O lost by evaporation and evapotranspiration. As the water is evaporated in the atmosphere, the salts are left behind in the soil. The process of accumulation of soluble salts in these soils is called salinization. The salts are mostly NaCl, Na_2SO_4, $CaCO_3$, and $MgCO_3$. In the past, the soils developed were called *saline soils, white alkali soils*, or *solonchaks*. They belong to the zonal-type soils. Salinization can also occur locally and develops the intrazonal type of saline soils (e.g., soils reclaimed from the sea bottom and soils in coastal areas affected by the tide).

Sodication and Alkalinization

The addition of salts to the soil may result in saturating the soils exchange complex with Na. The process of progressively increasing the Na saturation of the soils exchange complex is called *sodication*. The soils formed are called *sodic soils, solods, solonetz,* or *black alkali soils*. If these soils occur only in small areas (in small localized spots), they are often called *slick spots*.

Kamphorst and Bolt (1976) reported that sodication does not necessarily yield a rise in soil pH. Many sodic soils are neutral in reaction, whereas a number of solonetzic soils are even acid in reaction. In sodic soils with a neutral soil reaction, the Na salts are neutral salts such as NaCl.

The strong alkaline reaction (pH = 10) of most sodic soils is caused by alkalinization. The latter is due to hydrolysis of Na^+ ions or Na_2CO_3 compounds:

$$Na_2CO_3 + 2H_2O \rightleftharpoons 2Na^+ + 2OH^- + H_2CO_3$$

The OH^- ions produced will increase the soil pH, whereas the Na^+ is saturating the exchange complex. The latter, in turn, may undergo hydrolysis, which also contributes toward increasing the OH^- ion concentration in the soil:

$$\boxed{Clay}\ Na + H_2O \rightleftharpoons \boxed{Clay}\ H + Na^+ + OH^-$$

9.13 CHEMICAL CHARACTERIZATION OF SALINE AND SODIC SOILS

It is apparent from the foregoing that soil pH is not a good method for characterization of these soils. The saline soils have a soil pH = 8.5 or lower. The sodic soils may possess a soil pH = 10, but some of the soils may be neutral, whereas others are acid in reaction. To distinguish saline and sodic soils from other soils, the U.S. Salinity Laboratory (Richards, 1954) proposed to use as criteria soluble salt and exchangeable Na^+ content. These parameters are expressed in terms of (1) electrical conductivity (EC_e) for salt content and (2) exchangeable sodium percentage (ESP) for exchangeable Na^+ content. The salinity of the soil is measured by measuring the EC_e in millisiemens per centimeter (mmho/cm) of the soils saturated extract. The latter is obtained by suction and filtration of a water-saturated soil paste. The exchangeable sodium percentage is calculated using the formula as follows:

$$\text{ESP} = \frac{\text{Exchangeable Na}^+ \text{ Ions}}{\sum \text{Exchangeable Cations}} \times 100\%$$

Based on ESP and EC_e values, three groups of soils are recognized: (1) saline soils, (2) saline–alkali soils, and (3) nonsaline–alkali (sodic) soils. The saline soils are characterized by $EC_e > 4$ mmho/cm at 25°C, and ESP $< 15\%$. Dispersion of saline soils occurs at ESP $= 15\%$. The soil pH is ordinarily less than 8.5. Because of the presence of excess salts and low amounts of Na$^+$ in exchange position, these soils are usually in a flocculated state, and their permeability is considered to be equal or higher than the two other soils. The saline–alkali soils are soils with $EC_e > 4$ mmho/cm at 25°C, and ESP $> 15\%$. These soils have both free salts and exchangeable Na$^+$. As long as excess salts are present, the soil is flocculated and the pH is normally ≤8.5. When the soils are leached, the free salt content decreases, and the soil reaction may become strongly alkaline (pH > 8.5) because of hydrolysis of the exchangeable Na$^+$. Nonsaline–alkali soils are characterized by $EC_e < 4$ mmho/cm at 25°C, and ESP $> 15\%$. Most of the Na$^+$ is in exchangeable form, and very small amounts of free salts are present in the soil solution. The soil pH ranges from 8.5 to 10.0. As a result of irrigation, strong alkaline conditions may develop in these soils and pH values reaching 10 are common.

The selection of the critical value for EC_e at 4 mmho/cm was reported to be based on the expected salt damage to crops. The EC_e value of 4 mmho/cm originated with Scofield in 1942, who considered the soil to be saline at 4 mmho/cm or above. At the latter values the yield of many crops is restricted. Kamphorst and Bolt (1976) indicated that an EC_e of 4 mmho/cm corresponded to an osmotic pressure at field capacity of 5 bars. At EC_e values between 2 and 4 mmho/cm, only very sensitive crops will be affected, and at values below 2 mmho/cm the effect of salinity is negligibly small (Figure 9.2). The decision to use an ESP value of 15% is very arbitrary, since no sharp changes in soil properties have been observed as the degree of saturation of the exchange complex with Na$^+$ ions is increased. Moreover, different crops will react differently to the same ESP value (Kamphorst and Bolt, 1976; Richards, 1954). The U.S. Salinity

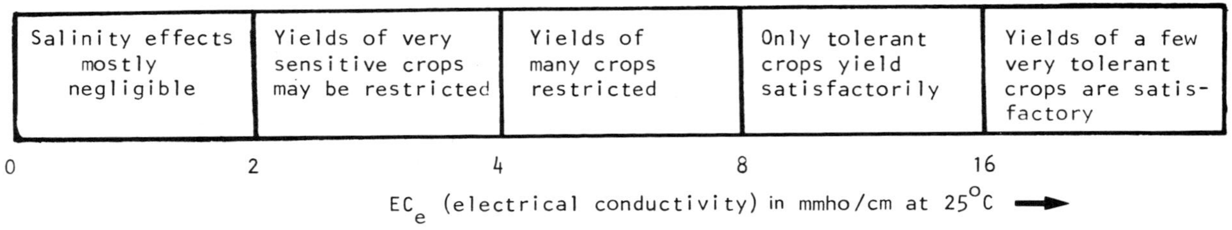

Salinity effects mostly negligible	Yields of very sensitive crops may be restricted	Yields of many crops restricted	Only tolerant crops yield satisfactorily	Yields of a few very tolerant crops are satisfactory

0 2 4 8 16 →

EC_e (electrical conductivity) in mmho/cm at $25°C$

Figure 9.2 The effect of degree of salinity, as expressed in EC_e values, on yields of crops according to the U.S. Salinity Laboratory. (From Richards, 1954.)

Laboratory has used, from history and experience, the ESP = 15% as a boundary limit to distinguish nonalkali from alkali soils.

9.14 EFFECT OF SALINIZATION AND SODICATION ON PLANT GROWTH

The accumulation of soluble salts in soils severely inhibits plant growth. It induces plasmolysis (see Sect. 6.6), by which H_2O moves out of the plant into the soil solution.

The presence of high amounts of Na^+ ions may keep the soil particles suspended. Upon drying, the soil may cake, and crust formation develops at the surface. The latter decreases soil porosity and aeration is severely inhibited.

The high pH in many of the soils also reduces availability of many micronutrients. These soils frequently encounter Fe, Cu, Zn, or Mn deficiencies.

9.15 IRRIGATION OF SALINE AND SODIC SOILS

Reclamation and management of the saline and sodic soils are based mainly on proper irrigation and drainage, on the exchange of Na^+ for Ca^{2+} on the exchange complex, and on the use of salt-tolerant crops.

Salinity Hazard

To make saline soils arable, leaching of excess salts by irrigation is usually conducted. A proper drainage method and the use of irrigation water with the proper salt quality are necessary. In this respect, the electrical conductivity EC_e is frequently used as an index for salinization hazard. The hazard of salinization is considered low if the irrigation water used has an $EC_e < 0.75$ mmho/cm (Richards, 1954; Taylor and Ashcroft, 1972).

Salinity hazard	EC_e (mmho/cm at 25°C)
Low	<0.75
Medium	0.75–1.5
High	1.5–3.0
Very high	>3.0

In arid regions, salinization is a natural phenomenon. Therefore, the chances for salinization are considered very high if water with an $EC_e = 3.0$ or higher is used for irrigation over many years, even on nonsaline soils.

Hazard of Sodication

The hazard of sodication is usually estimated by the use of the sodium adsorption ratio (SAR). The SAR formula is as follows (Richards, 1954):

$$SAR = \frac{(Na^+)}{\sqrt{(Ca + Mg)/2}} \qquad (9.7)$$

The concentration of Na^+ and $Ca^{2+} + Mg^{2+}$ can be expressed in millimoles per liter (Kamphorst and Bolt, 1976) or in milliequivalents per liter (Taylor and Ashcroft, 1972). Since the sodic soils are highly saturated with Na^+ ions, it is necessary to use irrigation water with low SAR values on these soils. It is sometimes suggested to add gypsum to the irrigation water. But most often the gypsum is plowed under in the soil. This may ensure the development of low SAR values of the water in the soil. The Ca^{2+} may, at the same time, replace Na^+ from the exchange complex. Theoretically, any soluble Ca compound, that will not affect soil pH, can be used together with irrigation water to reduce the SAR value and exchange the Na^+ ions.

9.16 SALT BALANCE AND LEACHING RATIO

Irrigation will sometimes only wet the soil. This is a potential danger for salt buildup. In the management of saline and sodic soils, the so-called salt balance is taken into account. The latter means that the amount of salt brought into the soil must equal the amount of salt leached out of the soil. Therefore, more water must be applied over that needed to wet the soil. The additional water, used for leaching, is called the *leaching requirement* (LR) (Bernstein and Francois, 1973):

$$LR = \frac{EC_{iw}}{EC_{dw}}$$

where

EC_{iw} = electrical conductivity of irrigation water
EC_{dw} = electrical conductivity of saturation extract of saline soil
          which exhibited 50% decrease in yield

If the irrigation water increases in salinity over the years, the value of LR becomes larger. Bernstein and Francois (1973) suggested managing irrigation in such a way that the major water needs of crops are supplied at the minimum salinity level of the irrigation water.

9.17 IRRIGATION-INDUCED SALINIZATION AND SODICATION

The hazard of salinization and sodication is perhaps not limited to semiarid and arid region soils. Currently, it is common practice to also use supplementary irrigation in areas with measurable rainy seasons. With the easy access of water from the huge underground aquifers in the southern coastal plain region of the United States, large areas are now continuously being irrigated by the center pivot sprinkle system. No adequate disposal systems of the used irrigation water have yet been devised. The irrigation water reaching the soil is allowed to percolate naturally through the soil and return to the ground water in a more concentrated condition. A large part of the irrigation water may, perhaps, also evaporate leaving the salts in the surface soil. No investigations have yet been done on the potential hazard in salinization and sodication by the use of this water. The danger of salinization and sodication is somewhat reduced by the presence of a humid climate. However, over many years one can expect a reduction in quality of this irrigation water by the use of high amounts of fertilizers and in the absence of a proper drainage and disposal system.

10

Soil Chemistry and Soil Formation

10.1 WEATHERING PROCESSES

Weathering refers to the disintegration and alteration of rocks and minerals by physical and chemical processes. Physical weathering is caused by physical stresses within the rock or mineral. It causes the rocks to disintegrate into smaller-sized material, without changing the chemical composition. Chemical weathering is caused by chemical reactions, and definite chemical changes occur in the weathering products. For an illustration of the chemical reactions involved (e.g., solution, hydration, hydrolysis, oxidation, reduction, and carbonation) reference is made to the textbooks on principles of soil science (Brady, 1974, 1990; Donahue et al., 1977; Foth and Turk, 1978).

In nature, both physical and chemical weathering may occur simultaneously. Both usually precede soil formation from solid rocks. Although, by nature, physical weathering is of more importance at or near the soil surface, occasionally it may take place below the soil surface. Plant roots may contribute to physical weathering below the soil surface. By growing into cracks, they may rupture the rocks apart.

Chemical weathering can occur at the soil surface, in the solum, or below the solum (in the parent material). Therefore, Jackson and Sherman (1953) suggested distinguishing it into pedochemical and geochemical weathering. *Pedochemical* weathering refers to chemical weathering within the solum, whereas *geochemical* weathering is weathering below the solum. Essentially, a sharp separation between pedochemical and geochemical weathering, as reported by Buol et al. (1973), is difficult to realize in nature. The main chemical reactions, such as solution, hydrolysis, hydration, oxidation, reduction, and carbonation, take place in the solum as well as in the parent material. Leaching of K from micas, alteration of clays by H^+ ions, and interlayering and formation of clays can occur as a pedochemical or as a geochemical process. Regardless of this difference in opinion, weathering in general results in a decrease in particle size of materials, in the release of soluble material, and in the synthesis of new materials (clays and humus).

10.2 STABILITY AND WEATHERING OF SOIL MINERALS

Crystal Chemistry and Mineral Properties

The breakdown and stability of minerals are quite complex and require a complete understanding of crystal chemistry. The relative resistance of a mineral to weathering processes is determined by its internal structure. The latter depends on the strength of the atoms or ions binding their neighboring ions in the crystal lattice of the mineral. Four major types of binding forces between atoms in crystals have been reported (Evans, 1939): ionic, homopolar, metallic, and van der Waals forces. Although in many cases the structural bonds arising from interionic interactions cannot be attributed to one bond type, the bonds in the crystal structure of soil minerals are considered to be mostly ionic.

Several mineral properties are affected by the respective bond types (Table 10.1). The ionic and homopolar bonds between atoms yield hard crystals, with high melting points. On the other hand, van der Waals attraction gives rise to only weak bonds and relatively soft crystals, with low melting points.

Table 10.1   Selected Physical and Structural Properties of Minerals as Related to Bonding Type

Mineral property	Type of bonds			
	Ionic	Homopolar	Metallic	van der Waals
Mechanical	Stong, hard	Strong, hard	Variable	Weak, soft
Thermal	High melting point	High melting point	Variable melting point	Low melting point
	Low thermal expansion	Low thermal expansion		High thermal expansion
Electrical	Nonconducting	Nonconducting	Conducting	Nonconducting
Optical	Variable	High refractive index	Opaque	Transparent
Structural	High coordination	Low coordination	Very high coordination	Very high coordination
	Moderately high density	Low density	High density	

Source: Evans (1939).

Coordination Theory and Pauling's Rules

The structure of soil minerals is formed by regular groupings of anions packed closely around a cation. Since most soil minerals are in oxide forms, the anions are usually oxygen atoms. The number of anions surrounding the cation is called the *coordination number* and is dependent on the respective ionic radii. A cation surrounded by three anions in an equilateral triangular configuration has a coordination number of 3. An arrangement of four anions around the cation is called a fourfold coordination. Such a crystal structure is called a *tetrahedron*. Silica tetrahedrons are examples of fourfold coordination structures. A sixfold coordination structure is a configuration with six anions around the cation, yielding an *octahedron*, such as the aluminum octahedron. With ions of larger dimensions, coordination numbers of 7–12 are possible.

This coordination theory in crystal structure is slightly different from the one used in complex compounds (chelates) formed by coordination reactions. However, viewed from the standpoint of broad generalities, some similarities between the two types of coordination theories are present. Complex formation and chelation by coordination reactions will be discussed in Chapter 11.

Since the atomic bonds in many of the soil minerals are ionic, the crystal structure of these minerals obeys the principles of ionic crystals as formulated by Pauling's rules (1929). The first rule states that "a coordinated polyhedron of anions is formed about each cation with the cation–anion distance being determined by the sum of the radii, and the coordination number of the cation being determined by the radius ratio." Not only are crystal structure and associated coordination number dependent on the size of the ions, they must, at the same time, obey the law of electroneutrality. A large cation, therefore, can coordinate more anions around it and still keep the anions apart. Smaller cations are capable of coordinating fewer anions. The limiting factor for each crystal arrangement is formulated by the radius ratio of the ions involved. The radius ratio is expressed as follows:

$$\text{Radius ratio} = \frac{r_c}{r_a}$$

where

r_c = radius of cation
r_a = radius of anion

A minimum radius ratio exists for each coordination number (Table 10.2). If the radius ratio is below a minimum value, the cation can coordinate only the next smaller number of anions. For example, a limit range of the radius ratio of 0.155–0.255 indicates that the cation in any cation–anion combination with a radius ratio r_c/r_a of 0.155 is capable of being in close association with only three anions. It will coordinate four anions if the radius ratio satisfies the value of 0.255 or larger.

Table 10.2 Relationship of Radius Ratio r_c/r_a and Coordination Number

Limit range of r_c/r_a	Coordination number	Crystal geometry
0–~	1	
0–~	2	Angular
0.155–0.255	3	Trigonal planar
0.255–0.414	4	Tetrahedron
0.414–0.732	6	Octahedron
0.732–1.0	8	

It should be noted again that the values in Table 10.2 are the ranges within which the radius ratio of atoms arranged geometrically in a crystal can vary without affecting the corresponding coordination number. The absolute values of r_c/r_a ratios between any two atom pairs are generally smaller than those stated in Table 10.2. For example, the absolute radius ratio of silicon and oxygen $r_{Si}/r_O = 0.42/1.40 = 0.300$ qualifies for a fourfold coordination. However, the absolute radius ratio between aluminum and oxygen is $r_{Al}/r_O = 0.51/1.40 = 0.364$. The latter does not fall within the limit range for a sixfold coordination.

Pauling's second rule states that "in a stable coordination structure, the total strength of the valency bonds which reach an anion from all neighboring cations equals the charge of the anion." This rule, also known as the *electrostatic valency principle*, indicates that the charge of the cation must be shared (divided) equally by the number of bonds to the neighboring anions. At the same time, the number of these bonds depends upon the coordination number of the cation. This rule provides for a symmetric arrangement of bonds of equal strength around each cation. Applied to soil minerals, Pauling's rule results in the following arrangement of charges. In soil minerals, mostly oxides, the number of oxygen atoms normally packed around the cation is considered the coordination number. The coordination number of silicon is 4, that of aluminum may be either 4 or 6, whereas iron and magnesium have coordination numbers of 6. Potassium is a large ion and is characterized by a coordination number of 12. Since each silicon ion in the tetrahedron is surrounded by four oxygen atoms, Pauling's rule says that the bond strength is equally shared among the tetrahedral bonds; in other words, the charge of the silicon ion is divided by the number of bonds (by 4). Consequently, each oxygen atom in the tetrahedron has half of its charge satisfied by the silicon to which it is bonded (Figure 10.1). The remaining unsatisfied valencies of the oxygen atoms are balanced by association with another silicon or with two Al^{3+} or Mg^{2+} ions. The sharing of an oxygen atom by two silicon ions develops the basic silicate mineral structure.

In an aluminum octahedron, the Al^{3+} ion is surrounded by six oxygen atoms. The bond strength with each oxygen atom contributed by the Al^{3+} ion is, therefore, $3/6$ or $1/2$. Applying Pauling's second rule in a similar manner, Fe^{2+} or Mg^{2+} in octahedral position will contribute only $2/6 = 1/3$ charge to each of the oxygen atoms.

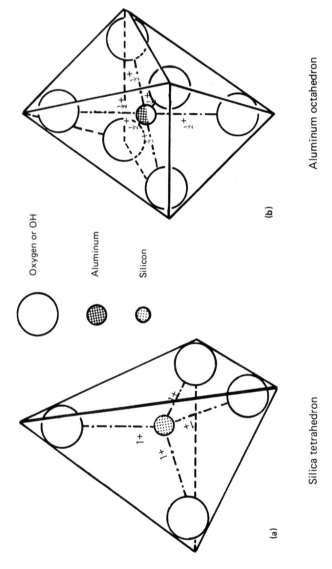

Silica tetrahedron   Aluminum octahedron

Figure 10.1   (a) The charge of the silicon atom is shared equally with the four surrounding oxygen atoms; consequently, each oxygen has only half of its charge satisfied by the silicon. (b) The charge of the aluminum atom is shared equally with the six surrounding oxygen atoms; therefore, the bond strength contributed by the Al^{3+} ion to each oxygen atom is only $\frac{1}{2}$.

Stability of Minerals and Bond Strength

As discussed previously, the fundamental units of silicate minerals are SiO_4 tetrahedrons. The latter can be joined together in several ways, and, depending on the arrangement of these SiO_4 tetrahedra, the minerals have been distinguished into cyclo-, ino-, neso-, phyllo-, soro-, and tectosilicates (see Sect. 5.1). Single or several units of tetrahedra can be linked together by other cations in the mineral framework. For example, double chains of silica tetrahedra can be linked together by Ca and Mg atoms, such as in amphiboles; or SiO_4 and AlO_4 tetrahedra are linked together by alkali or alkaline earth metals located in the lattice interstices, such as in feldspars. Whatever the structural arrangement is, the Si—O—Si linkage, called the *siloxane linkage* (Sticher and Bach, 1966) requires the greatest energy to form, compared with the other cation–oxygen bonds (Table 10.3). The data in Table 10.3 show that Si—O bonds are the strongest bonds, requiring 3110–3142 kg cal/mol for their formation. Aluminum–oxygen bonds are the next strongest (1793–1878 kg cal/mol needed for formation), and the bonds between the other metal ions and oxygen appear to be the weakest (299–919 kg cal/mol). The greater the number of Si—O bonds by linkage of increasingly larger

Table 10.3   Energies of Formation of Cation–Oxygen Bonds

Cation	Energy of formation (kg cal/mol)
Si^{4+} (nesosilicates)	3142
Si^{4+} (inosilicates, single chain)	3131
Si^{4+} (inosilicates, double chain)	3127
Si^{4+} (phyllosilicates)	3123
Si^{4+} (tektosilicates)	3110
Al^{3+} (framework)	1878
Al^{3+} (nonframework)	1793
Fe^{3+}	919
Mg^{2+}	912
Ca^{2+}	839
H^+ (in OH)	515
Na^+	322
K^+	299

Source: Paton (1978) and Keller (1954).

numbers of silica tetrahedra through oxygen sharing, the greater will become the resistance to weathering. On the basis of a progressive increase of oxygen sharing between adjacent silica tetrahedra, Keller (1954) and Birkeland (1974) ranked the silicate groups as follows: nesosilicates < inosilicates (single chain) < inosilicates (double chain) < phyllosilicates < tectosilicates.

To correlate such a ranking with a corresponding increase of resistance against weathering is subject to question. For example, the clay minerals belonging to the phyllosilicates are more resistant to weathering than feldspar and leucite, which are tectosilicates. However, in terms of comparison between biotite (phyllosilicates) and feldspar (tectosilicate), the foregoing ranking may have some value. Apparently, susceptibility to weathering is not affected by only the mineral structure, but a number of additional factors may also play an important role in mineral breakdown, as will be discussed with the individual silicate groups.

If the foregoing assumption is valid, stating that bonds requiring the greatest energy to form will also be the most resistant to attack by weathering, the data in Table 10.3 suggest that destruction of nonframework cation–oxygen bonds, such as Mg—O and Ca—O, will be relatively easier than the decomposition of the siloxane bonds, considered to be the silicate framework bonds. Cleavage of the siloxane bonds is made possible especially by interaction with chelating substances (Sticher and Bach, 1966). After collapse of the nonframework bonds, the tetrahedra may begin to break down when aluminum is present in the tetrahedral position. Since the weaker bonds are subject to attack first, the implication is that energy requirements for weathering may be considerably less than energy requirements for formation of the bonds.

In view of these considerations, the minerals in the various silicate groups are expected to differ in the way they respond to attack by weathering.

Cyclosilicates

The structure of this group is characterized by six-membered hexagonal rings of silica tetrahedra linked together by cations, such as Mg, Na, or Fe. The bonds formed by the latter are the weakest spots, but because of the abundance of Si—O linkages, the minerals in this group are considered relatively stable.

Inosilicates

The inosilicate group has in its structure single-chain (pyroxenes) and double-chain (amphiboles) silica tetrahedra linked together by Ca, Mg, or Fe. Because of the presence of many weak spots provided by the Ca—O, Mg—O, or Fe—O bonds, these minerals tend to weather rapidly.

Nesosilicates

The minerals in this group are composed of single tetrahedra linked together by Mg^{2+} and Fe^{2+} ions. To effect a breakdown, it is considered sufficient to sever the weaker Mg—O or Fe—O bonds. Notwithstanding the bond energy considerations, susceptibility of the minerals in this group to breakdown by weathering appears to vary considerably from one mineral to another (e.g., olivine versus zircon). The tight packing of oxygen atoms known to exist in zircon makes this mineral comparatively hard. On the other hand, the looser packing of oxygens in olivine makes the mineral weather faster.

Phyllosilicates

Linkages of silica tetrahedra and aluminum octahedra sheets by mutually shared oxygen atoms form the basis for the structure of this group. Some of the minerals (e.g., biotite and muscovite) are relatively susceptible to weathering; others like the clay minerals are resistant weathering products and further breakdown of clays is difficult. Disruption of the mineral usually occurs through removal (or replacement with OH) of interlayer ions or through cleavage of Al—O bonds in tetrahedral and octahedral positions.

Sorosilicates

Individual and linked silica tetrahedra formed by mutually shared oxygen are the basis for the structure of this group. Consequently, they are rather difficult to decompose. However, decomposition may take place in tetrahedra in which Al has substituted for Si.

Tektosilicates

The minerals are considered solid solution minerals with a framework of silica tetrahedra, in which the cavities are occupied by Na, Ca, and so on. The minerals in this group may also vary considerably in their resistance to weathering (e.g., leucite and plagioclase versus potash feldspars). The relative degree of close packing of atoms in their structural framework may be the reason for such a variability in weathering. Increased substitution of Al for Si in tetrahedra of plagioclase minerals is also considered a factor that makes this mineral weaker than potash feldspar.

10.3 WEATHERING OF FELDSPARS AND THE SILICA POTENTIAL

An important process in chemical weathering is the decomposition of soil minerals by hydrolysis. This can be illustrated by the decomposition reaction of orthoclase:

$$2KAlSi_3O_8(c) + 2H^+(aq) + 9H_2O(1) \rightleftharpoons H_4Al_2Si_2O_9(c) \\ + 4H_4SiO_4(aq) + 2K^+(aq) \tag{10.1}$$

By assuming the activities of orthoclase, water, and kaolinite unity, the mass action law gives

$$K = \frac{(H_4SiO_4)^2(K^+)}{(H^+)}$$

in which K denotes the decomposition or equilibrium constant. By taking the log, this equation becomes

$$\log K = \log \frac{(K^+)}{(H^+)} + 2 \log(H_4SiO_4) \tag{10.2}$$

In p ($-\log$) form, Eq. (10.2) is considered the chemical potential determining weathering stability of orthoclase. If the activity of H_4SiO_4 decreases below the value of the silica potential (pH_4SiO_4) of quartz or amorphous silica, orthoclase will decompose and form kaolinite as indicated by reaction (10.1). In equilibrium condition the ratio $(K^+)/(H^+)$ is unity (Garrels and Christ, 1965), and only the activity of H_4SiO_4 remains to control stability of orthoclase.

10.4   WEATHERING OF KAOLINITE AND THE GIBBSITE POTENTIAL

After formation, kaolinite is also subject to decomposition and can be transformed into gibbsite by weathering. The following reaction (Kittrick, 1967, 1969; Tan et al., 1973) determines the stability of kaolinite:

$$Al_2Si_2O_5(OH)_4(c) + 5H_2O(aq) \rightleftharpoons 2Al(OH)_3(c) + 2H_4SiO_4(aq)$$
(10.3)

In pure condition, kaolinite, gibbsite, and H_2O are considered unity; therefore, the following relationship follows from Eq. (10.3):

$$pK = 2pH_4SiO_4$$
(10.4)

The assumption is made that after part of the kaolinite is converted into gibbsite, the solution soon becomes saturated relative to gibbsite. At this point, kaolinite and gibbsite exist together in equilibrium, satisfying Eq. (10.3). Equation (10.4) is considered the chemical potential determining stability of kaolinite. Kaolinite is stable if pH_4SiO_4 is smaller than 4.73 (Tan et al., 1973).

Gibbsite, in turn, may be converted into Al^{3+}. The decomposition reaction can be written as

$$Al(OH)_3(c) + 3H^+(aq) \rightleftharpoons Al^{3+}(aq) + 3H_2O(l)$$

for which is valid

$$pH - \tfrac{1}{3}pAl^{3+} = 2.7$$
(10.5)

The expression $pH - \tfrac{1}{3}pAl^{3+}$ is called the *gibbsite potential*. Again, it can be argued that this chemical potential determines stability of gibbsite. Gibbsite is stable only if the gibbsite potential is less than 2.7 and when the activity of H_4SiO_4 is very small (Kittrick, 1967, 1969; Tan et al., 1973).

10.5   STABILITY AND PHASE RELATIONSHIPS OF SOIL MINERALS

After the stability of feldspar, kaolinite, gibbsite, and other soil minerals have been formulated as chemical potentials, a phase diagram

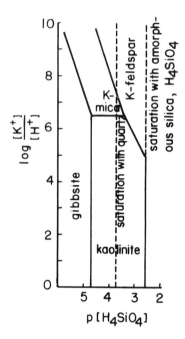

Figure 10.2 Stability and phase diagram of soil minerals.

can be drawn. This diagram delineates the regions at which the minerals are stable and shows the points at which they are not stable and start to decompose to form another mineral. An example of such a diagram is given in Figure 10.2.

The line bordering the areas of gibbsite and kaolinite is the chemical potential as expressed by $pH_4SiO_4 = 4.73$. All points on this line represent conditions at which both kaolinite and gibbsite coexist in equilibrium. In the gibbsite area (to the left of the line, $pH_4SiO_4 > 4.73$), only gibbsite is stable. If kaolinite is present at $pH_4SiO_4 > 4.73$, it will automatically decompose and form gibbsite. In the kaolinite area (to the right of the line, $pH_4SiO_4 < 4.73$), kaolinite is stable. If gibbsite is present at $pH_4SiO_4 < 4.73$, the presence of H_4SiO_4 will resilicate gibbsite into kaolinite. A similar discussion can be given for the other lines and areas.

Many types of stability diagrams can be made. Some are relatively simple, others are three-dimensional and very complex (Garrels and Christ, 1965). However, all of them have as a purpose the prediction of possible successive alteration of minerals with gradual changes in ion activity ratios as weathering proceeds.

10.6 EFFECT OF CHELATING AGENTS ON WEATHERING

Evidence has been reported that soil organic matter has a significant effect on weathering. Occasionally, the degree of weathering induced by soil organic matter may be more important than that brought about by chemical reactions alone.

By the decomposition of organic matter, several organic compounds are released or synthesized. Most of them, such as humic and fulvic acids, have the capacity to chelate or complex metal ions (Schnitzer and Kodama, 1976; Tan, 1976a, 1978a). Therefore, they may be able to pry loose Al and Fe from micas, feldspars, and kaolinite, or any other soil mineral, thereby accelerating the decomposition process. The organic chelating agent may perhaps react with an exposed cation, followed by the movement of the complex compound or chelate into solution. As a chelate, Al and other metals may be rendered soluble over a pH range in which it is insoluble as an ion. This is of importance in the formation of spodic horizons in Spodosols. Also, the H^+ ions produced during decomposition of organic matter may be adsorbed by silicate clays. Hydrogen ion saturation of clays, which results in their gradual decomposition, has been discussed earlier.

10.7 SOIL FORMATION PROCESSES

The process of soil formation is a complex biological and chemical problem and is usually difficult to describe with a single reaction. Reactions may occur simultaneously, or a sequence of reactions one after another are involved. Simonson (1959) stated that the soil pedon is formed by the combined effort of additions of inorganic and organic materials to the surface, transformation of compounds within the soil, vertical transfer of soil constituents within the soil, and removal of soil components from the soil.

The types of processes involved vary according to the conditions. However, it is not the purpose of this topic to list them in this chapter. For a list of possible processes of soil formation reference is made to Buol et al. (1973).

It is perhaps more important to discuss major soil-forming processes with general applicability to development of soil pedons, such as desilication and translocation of Al and Fe related in the formation of argillic, albic, spodic, and oxic horizons.

Desilication

Desilication is a process in which silica is released from soil silicates. Part of the released silica reacts with alumina to form clays, whereas the remainder is subject to leaching. Consequently, the soil exhibits a loss in silica content and at the same time, has a residual accumulation of stable weathering products, including sesquioxides.

A process such as desilication may occur in the tropics or in the temperate regions in the presence of sufficient amounts of moisture and the right temperature. Usually, it is more pronounced in the humid tropics. In the past, this process was known as laterization or ferralitization. The reactions can be illustrated with the decomposition of orthoclase into kaolinite [see Eq. (10.1)] and with the decomposition of kaolinite into gibbsite [see Eq. 10.3)]. If the soil is well drained and permeability is rapid, the activities of dissolved ions and H_4SiO_4 are kept low by leaching. The end product of weathering will then be gibbsite. Under poorly drained conditions and slow permeability, leaching is inhibited. The latter results in an increase in H_4SiO_4 activity and $(K^+)/(H^+)$ ratio, leading to formation of illites (Van Schuylenborgh, 1971).

The degree to which Si can be leached out of soil depends on its capacity to remain in solution. The solubility of silica is determined by the law of polymerization. Present in concentrations below 140 ppm SiO_2 (25°C), silica is found mainly in the form of monosilicic acid, $Si(OH)_4$, which is considered a true solution (Millot, 1970; Krauskopf, 1956). The solubility of this silica remains constant at 140 mg/L in the range of pH 2–9 (Figure 10.3), but at pH values above 9, solubility of silica increases rapidly. If the concentration of silica in the solution exceeds 140 ppm, polymerization of silica occurs, and usually a mixture of polymers and monomers of $Si(OH)_4$ is found in the soil solution. The polymers will be precipitated by

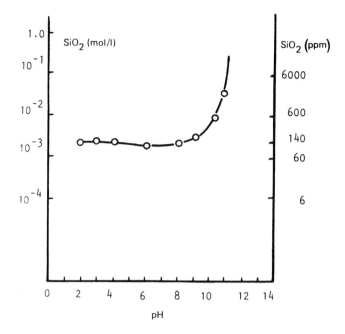

Figure 10.3 Solubility of silica as related to pH. (From Krauskopf, 1956.)

the introduction of small quantities of Al, or by decreasing pH (Paton, 1978), leaving the monomers in solution, which may tend to be leached out of the system.

Translocation of Clays

The process of clay translocation leads to enrichment of B horizons with clays. Such B horizons are called *argillic horizons* in the U.S. Soil Taxonomy. The clays have migrated from the A horizon because of an increase in peptization.

Evidence has been presented that clays and organic matter can form complexes (Greenland, 1971; Tan, 1976a). Although the exact mechanism is not yet known, the following hypothetical reaction serves as an example:

$$
\begin{array}{l}
\mid \\
-\,Si-O \\
\mid \qquad\qquad\searrow \\
\mid \qquad\qquad Al-OH + HO\!\!\left\langle\bigcirc\right\rangle\!\!COOH \;\rightarrow \\
-\,Si-O\qquad\nearrow \\
\mid
\end{array}
$$

CLAY   Org. comp

$$
\begin{array}{l}
\mid \\
-\,Si-O \\
\mid \qquad\qquad\searrow \\
\mid \qquad\qquad Al-O\!\!\left\langle\bigcirc\right\rangle\!\!COOH + H_2O \\
-\,Si-O\qquad\nearrow \\
\mid
\end{array}
$$

Clay-organic complex

This reaction adds an acidic group (COOH) to the clay surface and contributes a strong negative charge to the clay. The surface potential of the clay–organic complex is, therefore, larger than that of the clay alone. Consequently, the electrokinetic potential, formerly explained with the ζ potential, also becomes larger. This increases the capacity for peptization of clays. As a clay–organic complex, the clay remains suspended for a longer time and moves downward with the percolating water.

Several reactions are responsible for clay accumulation in the B horizon. The movement stops where the percolating water stops, and flocculation of clay may occur. Capillary withdrawal of water into the soil fabric deposits clay as clay skins on the walls of pores and peds.

Translocation of Aluminum and Iron

The downward movement of Al and Fe together with organic matter results in the formation of albic (E) and spodic (B_{hs}) horizons. This process was called *podzolization* in the past. It gives rise to the development of Spodosols (Podzols).

Most of the Fe subject to translocation comes from the decomposition of biotite and ferromagnesian minerals. The possible ionic forms of Fe(III) are Fe^{3+}, $(FeOH)^{2+}$, $Fe(OH)_2^+$, $Fe_2(OH)^{4+}$, and

$Fe(OH)_4^-$ (Van Schuylenborgh, 1966). The ionic forms of Fe(II) [i.e., Fe^{2+}, $(FeOH)^+$, and $Fe(OH)_3^-$ are less stable than those of Fe(III). Most of the soils in which translocation occurs are well drained. Therefore, most of the iron is in Fe(III) ionic form. The concentration of the Fe ions depends on the solubility of their respective solid phase.

Solubility constants of Fe compounds (Van Schuylenborgh, 1966)

$Fe(OH)_3^-$ amorphous $\rightleftharpoons Fe^{3+}$ $+ 3OH^-$	$pK = pFe + 3pOH = 38.2$
$eFe(OH)_3^-$ amorphous $\rightleftharpoons (FeOH)^{2+}$ $+ 2OH$	$pK = p(FeOH) + 2pOH = 26.3$
$Fe(OH)_3^-$ amorphous $\rightleftharpoons Fe(OH)_2^+$ $+ OH$	$pK = pFe(OH)_2 + pOH = 17.0$
Goethite	$pK = 45.2$
Lepidocrocite	$pK = 42.5$
Hematite	$pK = 42.5$
Maghemite	$pK = 41.0$

The assumption was made in the foregoing that in well-drained soils the iron is in Fe(III) form. However, whether Fe(III) or Fe(II) occurs under natural soil condition depends more precisely on the oxidation potential. If the following redox reaction of iron is studied:

$$Fe^{2+} \rightleftharpoons Fe^{3+} + e^-$$

then

$$k = \frac{[Fe^{3+}]}{[Fe^{2+}]}$$

Application of the Nernst equation gives the following relation:

$$E_h = E^0 + \frac{RT}{nF} \ln K$$

or

$$E_h = E^0 + 0.059 \log \frac{[Fe^{3+}]}{[Fe^{2+}]}$$

in which E_h is called the *oxidation potential*. After complete oxidation of Fe(II) to Fe(III), the concentration of Fe(II) ions becomes negligible small and can be neglected, so that the relationship changes to

$$E_h = E^0 + 0.059 \log[Fe^{3+}]$$

If, however, reduction processes prevail, the activity of Fe(II) becomes very large, so that for all practical purposes, activity of Fe(III) can be neglected. The oxidation potential assumes then the following relation:

$$E_h = E^0 - 0.059 \log \frac{[Fe^{2+}]}{[Fe^{3+}]}$$

Fe^{3+} negligible:

$$E_h = E^0 - 0.059 \log[Fe^{2+}]$$

Therefore, the oxidation potential E_h increases upon oxidation and decreases as a result of reduction processes. When the activity of Fe(III) equals that of Fe(II), the oxidation potential also equals the standard oxidation potential:

$$[Fe^{3+}] = [Fe^{2+}] \qquad E_h = E^0$$

With use of these oxidation potentials and pH values, several stability fields of Fe(II) and Fe(III) systems have been developed by various authors (Garrels and Christ, 1965; Hem and Cropper, 1959). An example is given in Figure 10.4.

From the diagram in Figure 10.4, it could be noticed that Fe(II) ions are stable at oxidation potentials below $E_h = 0.3$ if soil pH = 5–7. Only if the soil reaction is strongly acid, will Fe(II) ions remain stable at $E_h > 0.3$. The natural condition is represented by the shaded portion in Figure 10.4. A major part of this shaded area ($\frac{2}{3}$) lies in the stable field of ferric hydroxide, $Fe(OH)_2^+$, compounds. Hem and Cropper (1959) indicated that Fe(II) ions remain soluble at concentrations not to exceed 100 ppm ($E_h = 0.3$ and pH = 5.0). For a further discussion and detailed treatise on the principles of redox potentials see Chapter 2 and Section 10.8.

Almost all silicates are sources for Al. The ionic forms of Al(III) are Al^{3+}, $(AlOH)^2$, $Al(OH)_2^+$, $Al(OH)_4^-$, $Al_2(OH)_2^{4+}$, $Al_2(OH)_4^{2+}$, $Al_4(OH)_{10}^{2+}$, and $Al_6(OH)_{12}^{6+}$ (Van Schuylenborgh, 1966). As with

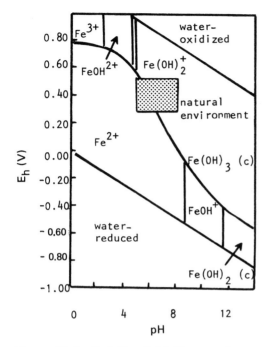

Figure 10.4 Stability field diagram for aqueous Fe(III)–Fe(II) systems. (From Hem and Cropper, 1959.)

iron, the concentration of the various Al ions is dependent on the solubility of their respective solid forms.

Solubility constants of Al compounds (Van Schuylenborgh, 1966)	
$Al(OH)_3$ amorphous $\rightleftharpoons Al^{3+}$ + $3OH^-$	$pK = pAl + 3pOH = 32.0$
$Al(OH)_3$ amorphous $\rightleftharpoons (AlOH)^{2+}$ + $2OH^-$	$pK = pAlOH + 2pOH = 23.4$
$Al(OH)_3$ amorphous $\rightleftharpoons Al(OH)_2^+$ + OH^-	$pK = pAl(OH)_2 + pOH = 14.1$
Gibbsite	$pK = 36.2\ (Al^{3+})$
Gibbsite (in H_2O)	$pK = 14.6\ (Al[OH]_4^-)$
Gibbsite (in base)	$pK = 0.57\ (Al[OH]_4^-)$

These data show that most of the solubility constants (pK values) of Al and Fe compounds are large. This means that the solubilities of the compounds are very low. The lower the pK value, the more soluble the compounds are in soils. It appears that only gibbsite in alkaline condition has a low pK value, perhaps sufficiently low to release some Al ions in solution. In addition to this, one must also take into consideration that the pH range in many soils is such that most Fe and Al compounds are essentially insoluble. Therefore, the possibility of migration of Al and Fe in the ionic forms shown is very small. Other agents are required to make Fe and Al more soluble.

Evidence has been presented showing that decomposition products of soil organic matter (Hodgson, 1969; Martin and Reeve, 1960a, b; Tan, 1978a) are capable of forming complexes with Fe and Al. As a complex, Fe and Al may remain soluble at pH ranges that make them usually insoluble. Another possibility is that the metal–organic complexes tend to disperse easily at low electrolyte concentration. In this way, they may move downward to deeper layers in the soil pedon. The loss of Fe, Al, and organic matter from the A horizon leads to the formation of an albic horizon.

Since the solubility constants of Al and Al–organic complexes are smaller than those of Fe and Fe–organic complexes, the Al complexes are more soluble than the Fe complexes. Consequently, Al complexes may move deeper in the soil pedon than the Fe complexes. In the B horizon, these compounds will be accumulated by (1) formation of insoluble complexes, (2) hydrolysis of metal complexes, (3) microbial attack of the organic ligands (Van Schuylenborgh, 1965), or (4) by a combination of these processes.

10.8 OXIDATION AND REDUCTION REACTIONS IN SOILS

Reduction and oxidation reactions occur in almost any soils. *Reduction* is, by definition, the gain of electrons, whereas *oxidation* is the loss of electrons. This can be illustrated by the following reaction:

$$Fe^{3+} + e^- \rightleftharpoons Fe^{2+} \tag{10.6}$$

Oxidation reactions are usually related to well-drained soil conditions. On the other hand, reduction processes are associated with poorly drained conditions, or where excess water is present. The latter develops gley formations.

Usually known as the *soil redox state*, this condition occurs in almost any soil. Both reduction and oxidation conditions can occur simultaneously in the pedon. While the surface layers of the pedon are in an oxidized state, the subsoil layers may be in a reduced condition owing to a fluctuating groundwater level. The latter may lead to pseudo-gley formation or to plinthization.

The redox condition of soils affects stability of iron and manganese compounds. To a certain extent, microbial activity and accumulation and decomposition of organic matter are also affected by the soils redox state. Fresh organic matter is thought to aid formation of reduced condition. Bloomfield (1951, 1953) reported that aqueous leaf extract reduced Fe(III) into Fe(II) in soils. In tidal floodwater areas, reduction processes play a considerable role in the formation of sulfur-rich soils.

Soils with different redox conditions may react differently upon N fertilization. In well-drained soils, ammonium N is subject to nitrification. However, if the ammonium fertilizer is applied to a reduced soil, such as to lowland rice or paddy soils, it remains available as ammonium.

Redox Potentials

The half-cell reaction for an oxidation–reduction system can be illustrated with Eq. (10.6) and attains the following general expression:

Oxidized state + e $\rightleftharpoons$ reduced state

The corresponding half-cell potential for this reaction obeys the Nernst equation:

$$E_h = E^0 + \frac{RT}{nF} \log \frac{\text{(oxidized state)}}{\text{(reduced state)}} \tag{10.7}$$

E_h is the redox potential; it is in fact the half-cell potential relative to a standard reference electrode. E^0 is a constant, called the *standard redox potential* of the system, and $RT/F = 0.0592$ at 25°C (see page 19; Garrels and Christ, 1965). If the activities of the oxidized and reduced species are unity, the ratio becomes 1, and the log equals

0. Consequently $E_h = E^0$. Therefore, the *standard redox potential* is defined as the redox potential of the system at which the activities of oxidized and reduced species are unity.

Application of Redox Potentials in Soils

The redox potential of soils varies with the reduced and oxidation state in soils. It is also associated with soil pH. E_h–pH relation are usually linear. (Garrels and Christ, 1965).

An illustration of the variations in redox potentials can be given when reduced iron is oxidized by an oxidizing agent (Figure 10.5). The curve in Figure 10.5 indicates that oxidation of Fe(II) causes the redox potential to rise. When 50% of Fe(III) is present, the redox potential equals 770 mV. This can be verified statistically by assuming that the half-cell reaction, $Fe^{3+} + e^- \rightarrow Fe^{2+}$, is characterized by a standard potential $E^0 = +770$ mV (see Table 2.1). The redox potential for this reaction has been formulated earlier as:

$$E_h = E^0 + 0.059 \log Fe^{3+}/Fe^{2+}$$

If

$$Fe^{2+} = Fe^{3+} (= 50\%)$$

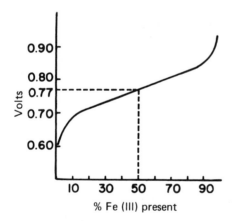

Figure 10.5 Redox potential curve of a waterlogged soil in the presence of oxidation of Fe(II) → Fe(III). (From Jeffery, 1960.)

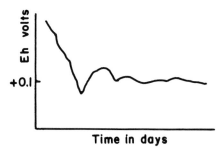

Figure 10.6 The redox potential curve as influenced by the length of flooding time. (From Jeffery, 1960.)

then:

$$E_h = E^0 + 0.059 \log 1$$

or

$$E_h = E^0 = 0.770V$$

From the foregoing, and from Eq. (9.5), it can be inferred that the redox potential of soils in an oxidized state is higher than that of soils in a reduced state. Jeffery (1960) reported an E_h value of -250 mV for soils in strong anaerobic condition. He also found that the redox potential was affected by flooding as shown in Figure 10.6. During the initial stage of flooding, the redox potential dropped rapidly, then it increased again and stabilized at approximately 100 mV. A system that has a stabilized redox potential is said to be *well poised*.

The redox potential can, therefore, be used to indicate the aeration status in soils:

Soil aeration status	E_h (V)
Well-aerated soils	0.4–0.7 or higher
Somewhat poorly aerated soils	0.3–0.35 or lower
Waterlogged soils	-0.4 or lower

Drastic changes take place in the physical, chemical, and biological conditions of soils with development of poor aeration or waterlogged condition. Oxygen content decreases rapidly in poorly aerated soils. Respiration of plant roots and microorganisms will rapidly consume the remaining oxygen in soil air. Therefore, oxygen content decreases with a decrease in E_h. At low E_h values, the dissolved oxygen in soil water will then be used by the microorganisms, and at very low E_h values even the combined oxygen, in the form of ferric oxides, nitrates, and sulfates, will be attacked.

A notable chemical effect resulting from waterlogging is the conversion of insoluble iron and manganese oxides into soluble Fe(II) and Mn(II), respectively. The reactions can be represented as follows:

$$Fe_2O_3 + 6H^+ + 2e^- \rightleftharpoons 2Fe^{2+} + 3H_2O$$
$$MnO_2 + 4H^+ + 2e^- \rightleftharpoons Mn^{2+} + 2H_2O$$

Consequently, in those poorly aerated soils in which reduction processes prevail, iron is found as Fe^{2+} iron, manganese as Mn^{2+}, nitrogen as NH_4^-, and sulfur as SO_3^- ion. On the other hand, these ions are present in the oxidized state in well-aerated soils (e.g., as Fe^{3+}, Mn^{4+}, NO_3^-, and SO_4^{2-}).

Biologically a decreasing oxygen content in soils produces drastic changes in the population of soil microorganisms. At low oxygen content, anaerobic microorganisms (*Actinomyces* sp.) may prevail over aerobic microorganisms. In medicine, the redox potential is applied for the detection of certain diseases. Low values of the redox potential around the gum increases the hazard for the occurrence of gum disease (gingivitis).

Stability of Iron Oxide and Hydroxides

The redox potential and pH are used to define stability relationships between iron oxide and iron hydroxide minerals. The formation of hematite from magnetite is considered an oxidation reaction. The reaction is simplified as follows:

$$2Fe_3O_4 + H_2O \rightleftharpoons 3Fe_2O_3 + 2H^+ + 2e^-$$

The redox potential of this system is then

$$E_h = E^0 + \frac{0.059}{2} \log \frac{(Fe_2O_3)^3(H^+)^2}{(Fe_3O_4)^2(H_2O)}$$

By assuming that H_2O and the mineral species are in a pure state, their activities are unity. Therefore, the equation changes to

$$E_h = E^0 + \frac{0.059}{2} \log(H^+)^2$$

or

$$E_h = E^0 - 0.059 \text{ pH} \tag{10.8}$$

Equation (10.8) is a linear relationship between E_h and pH. It indicates the boundary between stability of magnetite and hematite. If the redox potential of the system is larger than the E_h of Eq. (10.8), hematite is stable. If the redox potential of the system is smaller than the E_h formulated by Eq. (10.8), magnetite exists as the stable species.

Activity of Reduction Products

Van Breemen and Brinkman (1976) stated that flooding of aerobic soils reduced first the NO_3^- in soils. After the disappearance of nitrate, manganese would be reduced, followed by iron. The latter increased the concentrations of Mn^{2+} and Fe^{2+} ions during the initial period of reduced condition. The concentration of Mn^{2+} and Fe^{2+} ions decreased again upon continued flooding, and stabilized at a constant level. The net reaction is a condition in which Fe(III) and Fe(II) ions are present together. Such a condition is considered desirable for soils (Jeffery, 1960).

11

Chemistry of Soil–Organic Matter Interaction

11.1 COMPLEX FORMATION AND CHELATION

The terms *complex formation* and *chelation* have been used in soil science interchangeably. However, given the nature of bonding, a distinction can perhaps be made between complex formation and chelation. Complex formation is the reaction of a metal ion and ligands, through electron-pair sharing (Murmann, 1964; Mellor, 1964; Martell and Calvin, 1952). The resulting product is called a *metal coordination compound*. The metal ion is the electron-pair acceptor, and the ligand is the electron-pair donor. The metal ion serves as the central ion, and the organic ions are coordinated around it in a first coordination sphere.

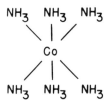

The number of ligands bonded to the central atom in a definite geometry is called the *coordination number*. Some of the organic ligands can bind the metal ion with more than one donor functional group. This type of bonding forms a heterocyclic ring, called a *chelate ring*. The process of formation of a chelate ring is called *chelation* (chelate means lobster claw). Hydrated metal ions in solution are also considered complexes with water, since they are surrounded by H_2O molecules (Perrin, 1964).

Almost any metal atom can serve as an acceptor atom, including K^+, Li^+, Na^+, Ag^+, and Au^+ (Murmann, 1964). A long-known complex compound with a monovalent ion is potassium ferrocyanide. A ligand can be an anion (Cl^- or $R—CH_2—COO^-$) or a neutral molecule (NH_3). The complexes produced can be cations, anions, or neutral molecules.

Several naturally occurring soil organic acids are capable of complexing metal ions. The reaction occurs mostly with the transition metals, Al, Fe, Cu, Zn, and Mn, and is often considered a special case of an adsorption process. Such kind of adsorption is quite different from the regular coulombic adsorption of cations in an electric double layer. The metal ion "adsorbed" in a complex reaction cannot be exchanged rapidly in the traditional manner of cation exchange reactions. The bonds in a complex compound are covalent bonds; hence, they are stronger bonds than the electrostatic bonds in cation exchange reactions. However, an exchange of the complexed metal is still possible, but such an exchange depends on many factors (e.g., soil pH, affinity of the metals for the ligand, and stability of the complexes). The exchange will take place more rapidly between transition metals. For example, an exchange of complexed Al by free Fe occurs easier than by Na^+ ions. The exchange by Na^+ is very difficult, since the Na^+ ion cannot occupy the center position of the Al^+ ion in the complex.

As will be discussed in Section 11.3, the chelation reactions have proved to be very beneficial in acid soils, in which the presence of large amounts of free Al, Fe, and Mn can create toxicity to plant growth (Tan and Binger, 1986). The organic acids can bind excessive metal ions in the soil and later release them to plants in smaller amounts, as needed. In this way Al toxicity can be controlled, because the organic acids prevent buildup of large amounts of micronutrients in the soil solution. Many of the organometal chelates formed are insoluble and prevent the metal from reaching the groundwater. The latter process is an important aspect in environmental quality.

The organic acids in soils of importance in complex reactions can be differentiated into two groups: nonhumified and humified organic acids. The *nonhumified organic acids* may range from simple aliphatic acids to complex aromatic and heterocyclic acids (e.g., formic acid, acetic acid, amino acid, benzoic acid, citric acid, oxalic acid, tannic acid, tartaric acid, and vanillic acid) (Tan, 1986). They have been released into the soil during the decomposition process of plant and animal residue, whereas others may have been released as root exudates. Some are intermediate products of plant and microbial metabolism, and others are the result of oxidative degradation of organic matter. For example, clover and pineapple plants contain large amounts of oxalic acids. Upon decomposition of the plants these acids enter the soil. The concentration of most of these acids in soil is generally very low. For example, the amount of acetic acid is between 0.7 and 1.0 mmol/100 g soil. The other group of acids, the *humified organic acids*, include fulvic acid (FA) and humic acid (HA; see Chapter 4). The concentrations of HA and FA are considerably higher than that of most of the nonhumified acids.

The effectiveness of these organic acids in complex formation and chelation reactions depends on their chemical reactivity. On the basis of chemical reactivity, the organic acids can be distinguished into two groups:

1. Organic acids in which the acidic characteristics are attributed to only the presence of carboxyl (—COOH) groups: These acids (e.g., formic, acetic, and oxalic acids) may exhibit some complexing capacity, but their main reaction is more through the acidic (H^+) effect or electrostatic attraction (see Figure 4.8).

2. Organic acids in which the acidic characteristics are attributed to the presence of both —COOH and phenolic-OH groups in their molecule (see Figure 4.7): These acids include humic and fulvic acids. They are capable of exerting a variety of reactions, including electrostatic bonding, coadsorption, complex formation, and chelation reactions. As indicated earlier, in the formation of chelates, the metal cation may connect itself to one or more radicals. The radicals involved may belong to the same or to different organic ligands. The maxi-

mum number of bonds is, however, governed by the coordination number of the cation.

Examples of two types of chelation reactions are presented here:

monodentate chelate
less stable

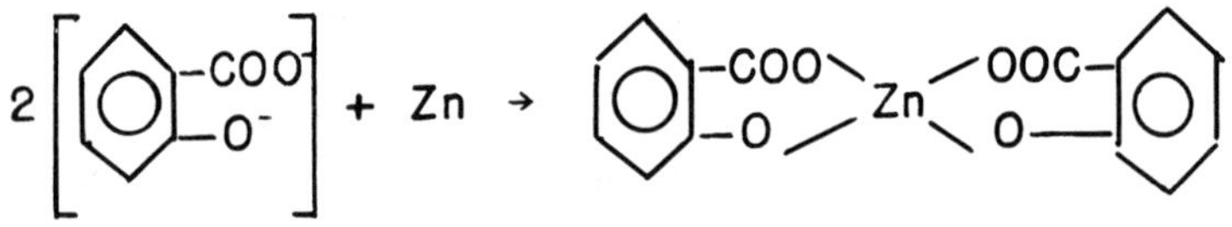

bidentate chelate
more stable

11.2 METAL–ORGANIC COMPLEX REACTIONS

Stability Constants

Several organic compounds including humic and fulvic acids (see pages 91–92) are capable of forming complexes with metal ions (Tan, 1978a; Stevenson, 1976a,b). Depending on the stability of the complexes, they can be soluble or insoluble in water. Assume that the following complex reaction occurs:

$$M^{2+} + 2HA \rightleftharpoons MA_2 + 2H^+$$

where

M = metal ion
HA = humic acid
MA_2 = metal–humic acid complex

then according to the mass action law, the equilibrium constant K is

$$K = \frac{(MA_2)(H^+)^2}{(M^{2+})(HA)^2}$$

By taking the log, this equation changes to

$$\log K = \log \frac{(MA_2)(H^+)^2}{(M^{2+})(HA)^2}$$

If the activities of HA and MA_2 are considered unity, then

$$\log K = \log \frac{(H^+)^2}{(M^{2+})}$$

or

$$\log K = 2 \log(H^+) - \log(M^{2+}) \tag{11.1}$$

where $\log K$ is called the *stability constant*. It determines the solubility of the metal complexes (Tan et al., 1971a,b). Tan et al. (1971b) have calculated $\log K$ values for metal–fulvic acid complexes. They found that the stability of the complex compound is high, if the value of $\log K$ is large. Therefore, the following data indicate that Cu–FA complexes are more difficultly soluble than Zn– or Mg–FA complexes. The degree of solubility is largest for the Mg–FA complex.

Stability constants of metal–fulvic acid complexes	Cu–FA	Zn–FA	Mg–FA
$\log K$ (pH 3.5)	7.15	5.40	3.42
$\log K$ (pH 5.5)	8.26	5.73	4.06

Clay–Organic Compound Complexes

Clay can also form complexes with organic compounds (Tan, 1976a; Theng, 1974, 1972; Greenland, 1971). The organic compounds can be cationic, anionic, and polar nonionic in nature.

Complex Formation With Organic Anions

Under ordinary conditions clay has a negative charge and, therefore, will repel organic anions. However, under certain conditions (see page 192) the broken edge surface of clay attains positive charges and will attract anions.

Negative adsorption is sometimes considered a possibility in this aspect (Theng, 1972). However, *negative adsorption* (see page 246) is defined as the repulsion of anions by negatively charged clay surfaces. Therefore, the relationship between negative adsorption and complex formation (coordination bonding) is rather obscure.

Several authors have noted the presence of negative adsorption of herbicides by montmorillonite (Bailey and White, 1970; Weber, 1970; Frissel and Bolt, 1962). Herbicides, such as 2,4-D and 2,4,5-T, are found to be negatively adsorbed by Na-montmorillonite in a medium with a pH above the pK_a value of the organic compound. Negative adsorption continues to be important, until the pH of the medium equals the pK_a, or the dissociation constant of the respective compound. Positive adsorption starts as soon as the pH is decreased below the pK_a. Bailey and White (1970) consider the dissociation constant to be a major factor in determining adsorption processes. The pK_a is used by these authors as an indicator for the degree of acidity or basicity of the substance. At pH values $>pK_a$, these organic compounds exist in the acidic form and therefore, behave as anions. As such, they are subject to attraction by positively charged clay surfaces, or by negative-anion adsorption. Positive charges are usually present at the edges of silicate clays, especially under strongly acidic conditions. Therefore, the main mechanism for adsorption of acidic organic compounds is negative adsorption. At pH values $<pK_a$, the compounds exist mainly in the basic form and, therefore, behave as cations. They will be attracted by negatively charged clay surfaces.

Another possibility of interaction between acidic organic compounds and silicate clays is through reactions with cations in exchange positions or through water molecules coordinated to these cations. These processes have been discussed earlier in Chapter 6.

Complex Formation With Organic Cations

Under certain conditions a number of organic compounds may be positively charged (e.g., amino compounds; see pages 73–74). Positive charges can also develop by the following processes. Mortland (1970) suggests that after adsorption by clay, the organic compounds become positively charged by accepting protons as follows:

1. H^+-saturated clays may donate the proton

$$R-NH_2 + H-clay \rightleftharpoons R-NH_3-clay$$

2. Water polarized by a cation can donate a proton to the organic compound:

$$(M \cdot mH_2O)^{m+} + NH_2-R \rightleftharpoons (NH_3-R)^+$$
$$+ [M \cdot (m - 1)H_2O]^{m-1}$$

3. By the presence of a protonated species that donates a proton to the organic molecule:

$$\underset{\substack{\text{protonated} \\ \text{species}}}{(HA)^+} = \underset{\substack{\text{organic} \\ \text{molecule}}}{NH_2-R} \rightleftharpoons A + (NH_3-R)^+$$

These organic compounds, having attained a positive charge, may replace inorganic cations on exchange positions or in interlayer surfaces of clays. Such an exchange follows the general laws of cation exchange reactions. The exchange occurs stoichiometricly and reaches a maximum equaling the CEC of clays (Hendricks, 1941; Greenland, 1965, 1971). If the organic cation occupies intermicellar spaces, analysis of basal spacings of clays indicates that the organic ion is adsorbed with its shortest axis perpendicular to the silicate layer (Theng, 1972).

Complex Formation With Amphoteric Organic Compounds

As discussed earlier, important organic substances in soils with amphoteric character are humic compounds, proteins, and amino acids. The presence of functional groups, such as carboxylic and amino groups, in their molecules gives them the ability to exist either as a cation, anion, or as a zwitterion (see page 74). The dominant ion species present in the soil solution depends on the soil reaction. The latter can be illustrated with the α-amino acid, L-alanine, as shown

$$
\underset{\substack{|\\H}}{\overset{\substack{NH_3^+\\|}}{CH_3-C-COOH}} \; \rightleftarrows \; \underset{\substack{|\\H}}{\overset{\substack{NH_3^+\\|}}{CH_3-C-COO^-}} \; \rightleftarrows \; \underset{\substack{|\\H}}{\overset{\substack{NH_2\\|}}{CH_3-C-COO^-}}
$$

Cation	Zwitterion	Anion
pH < isoelectric point	pH = isoelectric point	pH > isoelectric point

in the scheme above. The ion species in which the amino acid occurs governs the interaction with other soil components. In an acid soil reaction, or at pH values below the isoelectric point, amino acids are usually positively charged. As discussed on pages 73–74, the amino group can obtain an extra proton, and behave then as a cation. As such, the cationic form of amino acid can be attracted to the clay surface by cation exchange. The latter mechanism and proton transfer are expected to be the main processes for the interaction reactions at the clay–solution interface or in the intermicellar spaces of expanding layer silicates.

At neutral soil reactions, or soil pH close to the isoelectric point, amino acids are dipolar and behave as a zwitterion. Although most soils under field conditions are in this pH range, the pH at the clay interface is generally lower than the pH of the bulk solution. When present as a zwitterion, amino acid will interact through ion dipole reactions. The positive pole (NH_3^+) can be attracted directly to the negatively charged clay surfaces. On the other hand, the negative pole of amino acid (COO^-) can also undergo interactions with metal cations adsorbed on the clay surfaces.

In alkaline conditions, or soil pH above the isoelectric point, amino acids are negatively charged and possess anion characteristics. Although the anionic form is considered to be less important than the cationic and zwitterionic forms (Theng, 1974), it has the capability of reacting with positively charged clay surfaces, or it may be attracted to the surface of clays by cationic bridging.

Complex Formation with Nonionic Organic Compounds

The interaction between clay and uncharged organic molecules, such as alcohols and ethylene glycol, is made possible by the presence of exchangeable cations (Mortland, 1970). These cations are surrounded by water molecules arranged as a hydration shell. One of the water molecules can be exchanged for an organic ligand. The organic compound attached itself through the exchangeable cation to the clay surface. This reaction is also called *complex reaction*. If the cation has a low hydration energy, then the cation may form a direct coordinate bond with the oxygen atoms of the organic molecule.

The adsorption of organic molecules in intermicellar spaces is called *intercalation*. The basal spacing of clays increases stepwise as one, two, or three layers of organic compounds are intercalated. Compounds with strong polarity will orient themselves parallel to the silicate surface when intercalated (Hoffmann and Brindley, 1960).

Complex Formation and Mobility of Soil Constituents

Metal Mobility

The effect of chelation on metal mobility is caused by a change in ionic behavior. After chelation the cation is surrounded by the chelating ligands. The cation may be transformed into an anion. Anions will be repelled by negatively charge colloids. Therefore, they will remain mobile. The elements, once released by weathering in the soil, behave according to the conditions. Some will be used for clay mineral synthesis, such as Si, Al, and Fe, and others will be adsorbed by soil colloids. The latter can be made available to plant growth, or be translocated, later on, with the percolating water to deeper soil depth. They may also be removed from the soil profile. For some of the elements, such as K and Na, the latter is a simple process, since they easily form soluble substances. With Fe and Al, it is a complex problem. Iron and Al compounds are usually insoluble at the normal soil pH range. However, the solubility of these substances can be increased by complex formation or chelation of Fe and Al with soil humic compounds. Metals in the form of metal chelates can percolate with the rainwater downward in the soil profile. As a soluble chelate, the metal can also be taken up by plants by exchange reaction.

Complex formation can also occur with Ca^{2+} and Mg^{2+} ions. However, since the latter are relatively soluble, they can also exist as free ions. Consequently, the mobility of Ca^{2+} and Mg^{2+} is less dependent on complex formation than are Al^{3+} and Fe^{3+}.

Aluminum and Fe chelates are usually present in Spodosols, whereas Ca chelates occur mainly in Mollisols (Tan, 1978a). As explained earlier, Al and Fe chelates have high mobility and are the reasons for formation of albic and spodic horizons. On the other hand, the Ca chelates in Mollisols are immobile. This difference in mobility is due to the difference in organic ligands. In Spodosol, the organic ligands are mostly fulvic acids, whereas in Mollisols the ligands are mainly humic acid. Fulvic acid is the water-soluble humic compound and, consequently, may form soluble or easily dispersable chelates. Humic acids are insoluble in water and form stable Ca chelates in Mollisols.

Stability Diagrams of Metal Chelates

The cations in soils, such as H^+, Al^{3+}, Fe^{3+}, Zn^{2+}, Mn^{2+}, Ca^{2+}, and Mg^{2+}, may compete for bonding with the chelating agent. The

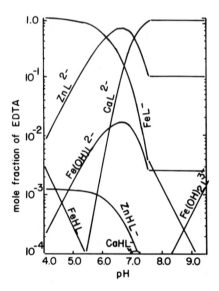

Figure 11.1 Stability diagram of Zn^{2+}, Fe^{3+}, Ca^{2+}, and H^+ complexes with EDTA and DTPA as influenced by soil pH. (From Lindsay and Norvell, 1969.)

cation that can form the most stable chelate in the particular condition will be bonded. Lindsay (1974) and Lindsay and Norvell (1969) have made a number of stability diagrams of Zn-, Fe-, Ca-EDTA. They reported that at pH 6–7, Zn-EDTA was stable. By the term *stable*, Lindsay and Norvell (1969) meant that the chelate existed as a soluble compound. Between pH 5.5 and pH 7.0, Fe-EDTA was the stable species in the chelate mixture (Figure 11.1). AT pH 7.0–7.3, Ca-EDTA was stable. These authors concluded that Fe^{3+} has displaced Zn^{2+} from Zn-EDTA at low pH, whereas Ca^{2+} has displaced Zn^{2+} or Fe^{3+} from Zn-EDTA or Fe-EDTA at high pH. Therefore, the presence of Fe at low pH may cause Fe to become chelated and remain soluble. On the other hand, Zn is then expelled from the chelate compound and precipitates as $Zn(OH)_2$. The latter may cause Zn deficiency.

Clay Mobility

The migration of clay in soils requires that clay remains in suspension. Dispersion and suspension are increased by (1) a low electrolyte or low base content in soils and (2) by the absence of positively charged colloids, or (3) by a combination of both of these factors. Soils high in $CaCO_3$ show little evidence of translocation of clays. However, clay can also form complexes with soil organic compounds. The latter increases its capacity to disperse (see pp. 293–294), and consequently increases its mobility.

11.3   COMPLEX FORMATION AND SOIL FERTILITY

Complex formation and chelation play an important role in improving soil fertility. In the preceding sections, it was shown that chelation increases the mobility and, consequently, the plant availability of many cations. The release of plant nutrients by weathering of soil minerals is usually a slow process. However, complex formation tends to accelerate the decomposition process of soil minerals and, accordingly, also accelerates the release of soluble nutrients.

The harmful effect of fixation of K and P is also offset by complex formation with organic ligands. Evidence has been shown by Tan (1978b) that humic and fulvic acids increase the release of K fixed in the intermicellar spaces of clays. It is expected that chelation or complex formation may also contribute toward making insoluble inorganic phosphates more soluble. The solubility of $AlPO_4$, $FePO_4$, or $Ca_3(PO_4)_2$ is expected to increase considerably by complex formation with humic compounds or other organic compounds. Humic and fulvic acids have a high affinity for Al, Fe, and Ca. Consequently, they will complete for these elements with the respective phosphates by complex formation, releasing in the process the phosphate ions in the soil solution.

Humic compounds are also effective in binding the micronutrients, such as Fe, Cu, Zn, and Mn. In acidic soils, these micronutrients are present in large amounts and cause toxicity problems to plants. By adding humus to acidic soils, some of the excess of micronutrients is taken out of the solution by complex formation with humic compounds. In time, they can be released again to plants in smaller amounts as needed. In this way, the chelate acts as a regulatory agent. From an environmental or ecological standpoint, complexing of heavy metal ions by humic compounds may temporarily reduce toxic hazards for human beings, animals, and plants. In some soils, the soluble fraction of the micronutrients Fe, Zn, Cu, and Mn can be deficient, since usually they are too insoluble. Chelation of these elements by soil organic matter increases their solubility (Lindsay, 1974). The latter helps in maintaining adequate levels of soluble micronutrients in the soil solution.

Appendix A

Fundamental Constants

Symbol	Name	Value
c	Velocity of light	2.9979×10^{10} cm/sec
e	Electonic charge	1.6021×10^{-19} C
L	Avogadro's number	6.0225×10^{23}
h	Max Planck's constant	6.6256×10^{-27} erg sec
F	Faraday constant	96.487 C/Eq
R	Gas constant	82.056 cm^3 atm/(mol)(deg)
		1.9872 cal/(mol)(deg)
		8.3143 J/(mol)(deg)
K	Kelvin temperature	$-273°C = 0$ K
k	Boltzmann constant	1.3805×10^{-16} erg/degree

Appendix B

Greek Alphabet

Greek letter		Greek name	Greek letter		Greek name
A	α	Alpha	N	ν	Nu
B	β	Beta	Ξ	ξ	Xi
Γ	γ	Gamma	O	o	Omicron
Δ	δ	Delta	Π	π	Pi
E	ε	Epsilon	P	ρ	Rho
Z	ζ	Zeta	Σ	σ	Sigma
H	η	Eta	T	τ	Tau
Θ	θ	Theta	Υ	υ	Upsilon
I	ι	Iota	Φ	φ	Phi
K	κ	Kappa	X	χ	Chi
Λ	λ	Lambda	Ψ	ψ	Psi
M	μ	Mu	Ω	ω	Omega

Appendix C

Periodic Classification of Elements

PERIODIC CLASSIFICATION OF ELEMENTS
based on $^{12}C = 12.0000$

Ia	2a	3b	4b	5b	6b	7b	8	8	8	1b	2b	3a	4a	5a	6a	7a	0
1 H 1.0080																	2 He 4.003
3 Li 6.939	4 Be 9.012											5 B 10.81	6 C 12.011	7 N 14.007	8 O 15.999	9 F 18.998	10 Ne 20.183
11 Na 22.990	12 Mg 24.31											13 Al 26.98	14 Si 28.09	15 P 30.974	16 S 32.064	17 Cl 35.453	18 Ar 39.948
19 K 39.102	20 Ca 40.08	21 Sc 44.96	22 Ti 47.90	23 V 50.94	24 Cr 52.00	25 Mn 54.94	26 Fe 55.85	27 Co 58.93	28 Ni 58.71	29 Cu 63.54	30 Zn 65.37	31 Ga 69.72	32 Ge 72.59	33 As 74.92	34 Se 78.96	35 Br 79.909	36 Kr 83.80
37 Rb 85.47	38 Sr 87.62	39 Y 88.91	40 Zr 91.22	41 Nb 92.91	42 Mo 95.94	43 Tc (99)	44 Ru 101.1	45 Rh 102.90	46 Pd 106.4	47 Ag 107.87	48 Cd 112.40	49 In 114.82	50 Sn 118.69	51 Sb 121.75	52 Te 127.60	53 I 126.90	54 Xe 131.30
55 Cs 132.91	56 Ba 137.34	57 La* 138.91	72 Hf 178.49	73 Ta 180.95	74 W 183.85	75 Re 186.2	76 Os 190.2	77 Ir 192.2	78 Pt 195.09	79 Au 197.0	80 Hg 200.59	81 Tl 204.37	82 Pb 207.19	83 Bi 208.98	84 Po (210)	85 At (210)	86 Rn (222)
87 Fr (223)	88 Ra 226.05	89 Ac** (227)															

Lanthanide * series													
58 Ce 140.12	59 Pr 140.91	60 Nd 144.24	61 Pm (147)	62 Sm 150.35	63 Eu 151.96	64 Gd 157.25	65 Tb 158.92	66 Dy 162.50	67 Ho 164.93	68 Er 167.26	69 Tm 168.93	70 Yb 173.04	71 Lu 174.97

Actinide ** series													
90 Th 232.04	91 Pa (231)	92 U 238.03	93 Np (237)	94 Pu (242)	95 Am (243)	96 Cm (247)	97 Bk (249)	98 Cf (251)	99 Es (254)	100 Fm (253)	101 Md (256)	102 No (254)	103 Lw (257)

Appendix D

X-Ray Diffraction 2θ d Spacing Conversion Table

2θ d Spacing Values for Cu Kα Radiation with λ = 1.5405 Å (0.1540 nm)

2θ	0.0	0.1	0.2	0.3	0.4	0.5	0.6	0.7	0.8	0.9
0	∞	882.63	441.32	294.21	220.66	176.53	147.11	126.09	110.33	98.076
1	88.263	80.245	73.555	67.897	63.047	58.845	55.167	51.922	49.038	46.457
2	44.135	42.033	40.122	38.378	36.779	35.308	33.950	32.693	31.526	30.440
3	29.425	28.476	27.587	26.751	25.964	25.223	24.522	23.859	23.232	22.636
4	22.071	21.532	21.020	20.531	20.065	19.619	19.193	18.785	18.394	18.018
5	17.659	17.312	16.979	16.660	16.352	16.054	15.768	15.491	15.225	14.967
6	14.717	14.476	14.243	14.017	13.798	13.586	13.381	13.181	12.988	12.800
7	12.617	12.440	12.267	12.099	11.936	11.777	11.622	11.471	11.325	11.182
8	11.042	10.906	10.773	10.644	10.517	10.394	10.273	10.155	10.040	9.9270
9	9.8168	9.7098	9.6042	9.5010	9.4001	9.3015	9.2053	9.1105	9.0173	8.9264
10	8.8378	8.7500	8.6645	8.5506	8.4989	8.4181	8.3387	8.2609	8.1847	8.1100
11	8.0360	7.9644	7.8935	7.8234	7.7549	7.6880	7.6220	7.5571	7.4932	7.4305
12	7.3688	7.3081	7.2484	7.1897	7.1320	7.0751	7.0192	6.9642	6.9100	6.8567
13	6.8042	6.7524	6.7015	6.6513	6.6019	6.5532	6.5053	6.4550	6.4114	6.3655
14	6.3203	6.2757	6.2317	6.1883	6.1456	6.1035	6.0619	6.0209	5.9804	5.9405
15	5.9011	5.8623	5.8239	5.7860	5.7488	5.7119	5.6755	5.6395	5.6041	5.5691
16	5.5345	5.5004	5.4666	5.4333	5.4004	5.3679	5.3358	5.3040	5.2727	5.2417
17	5.2111	5.1809	5.1510	5.1214	5.0922	5.0633	5.0348	5.0065	4.9787	4.9511
18	4.9238	4.8968	4.8701	4.8437	4.8176	4.7918	4.7663	4.7410	4.7160	4.6913
19	4.6669	4.6426	4.6187	4.5950	4.5715	4.5482	4.5253	4.5026	4.4801	4.4577

2θ d Spacing Values for Cu $K\alpha$ Radiation with $\lambda = 1.5405$ Å (0.1540 nm)

2θ	0.0	0.1	0.2	0.3	0.4	0.5	0.6	0.7	0.8	0.9
20	4.4357	4.4138	4.3922	4.3708	4.3496	4.3287	4.3079	4.2872	4.2669	4.2467
21	4.2267	4.2069	4.1872	4.1678	4.1486	4.1295	4.1106	4.0919	4.0733	4.0550
22	4.0367	4.0187	4.0008	3.9831	3.9656	3.9481	3.9309	3.9139	3.8969	3.8801
23	3.8635	3.8469	3.8306	3.8144	3.7983	3.7824	3.7666	3.7509	3.7354	3.7200
24	3.7047	3.6896	3.6746	3.6596	3.6449	3.6302	3.6157	3.6013	3.5870	3.5728
25	3.5587	3.5448	3.5309	3.5172	3.5036	3.4901	3.4767	3.4634	3.4502	3.4371
26	3.4241	3.4112	3.3984	3.3857	3.3731	3.3606	3.3482	3.3359	3.3236	3.3115
27	3.2995	3.2875	3.2758	3.2639	3.2522	3.2406	3.2291	3.2176	3.2063	3.1951
28	3.1839	3.1727	3.1617	3.1508	3.1399	3.1291	3.1184	3.1078	3.0973	3.0868
29	3.0763	3.0660	3.0557	3.0455	3.0354	3.0253	3.0153	3.0054	2.9955	2.9857
30	2.9760	2.9664	2.9568	2.9472	2.9377	2.9283	2.9190	2.9098	2.9005	2.8914
31	2.8823	2.8732	2.8643	2.8553	2.8465	2.8376	2.8289	2.8202	2.8116	2.8029
32	2.7945	2.7859	2.7775	2.7691	2.7608	2.7526	2.7443	2.7362	2.7281	2.7200
33	2.7120	2.7040	2.6961	2.6882	2.6804	2.6727	2.6649	2.6573	2.6496	2.6420
34	2.6345	2.6270	2.6195	2.6121	2.6048	2.5974	2.5902	2.5830	2.5757	2.5686
35	2.5615	2.5541	2.5474	2.5404	2.5334	2.5295	2.5196	2.5129	2.5060	2.4993
36	2.4926	2.4859	2.4793	2.4727	2.4661	2.4596	2.4531	2.4466	2.4402	2.4338
37	2.4274	2.4211	2.4149	2.4086	2.4024	2.3962	2.3901	2.3840	2.3779	2.3719
38	2.3659	2.3599	2.3540	2.3480	2.3421	2.3362	2.3305	2.3247	2.3189	2.3131
39	2.3074	2.3018	2.2962	2.2905	2.2849	2.2794	2.2739	2.2684	2.2629	2.2574

Appendix E

System International (SI) Units

SI unit	Symbol
Ampere (electrical current)	A
Candela (luminous intensity)	cd
Meter (length)	m
Mole (amount of substance)	mol
Kelvin (thermodynamic temperature)	K
Kilogram (mass)	kg
Second (time)	s
Square meter (area)	m^2

Factors for Converting into SI Units

U.S. unit	SI unit	To obtain SI unit multiply U.S. unit by
Acre	Hectare, ha	0.405
Acre	Square meter, m^2	4.05×10^3
Atmosphere	Megapascal, MPa	0.101
Calorie	Joule, J	4.19
Cubic foot	Liter, L	28.3
Cubic inch	Cubic meter, m^3	1.64×10^{-5}
Curie	Becquerel, Bq	3.7×10^{10}
Dyne	Newton, N	10^{-5}
Erg	Joule, J	10^{-7}
Foot	Meter, m	0.305
Gallon	Liter, L	3.78
Gallon per acre	Liter per ha	9.35
Inch	Centimeter, cm	2.54
Mile	Kilometer, km	1.61
Miles per hour	Meter per second	0.477
Ounce (weight)	Gram, g	28.4
Ounce (fluid)	Liter, L	2.96×10^{-2}
Pint	Liter, L	0.473
Pound	Gram, g	454
Pound per acre	Kilogram per ha	1.12
Pound per cubic foot	Kilogram per m^3	16.02
Pound per square foot	Pascal, Pa	47.9
Pound per square inch	Pascal, Pa	6.9×10^3
Quart	Liter, L	0.946
Square foot	Square meter, m^2	9.29×10^{-2}
Square inch	Square cm, cm^2	6.45
Square mile	Square kilometer, km^2	2.59
Ton	Kilogram, kg	907
Ton per acre	Megagram per ha	2.24

REFERENCES AND ADDITIONAL READINGS

Achard, F. K. (1786). Chemische Untersuchung des Torfs. Crell's Chem. Ann. 2:391–403.

Adler, H. H., P. F. Kerr, E. E. Bray, N. P. Stevens, J. M. Hunt, W. D. Keller, and E. E. Pickett (1950). Infrared spectra of reference clay minerals. Am. Petroleum Inst., Project 49. Clay Mineral Standards Preliminary Report No. 8. Cornell University, Ithaca, N.Y.

Ahmad, F., and K. H. Tan (1986). Effect of lime and organic matter on soybean seedlings grown in aluminum-toxic soil. Soil Sci. Soc. Am. J. 50:656–661.

Appelquist, T. W. (1986). Elementary particle. In McGraw-Hill Yearbook of Science and Technology, McGraw-Hill, New York, pp. 184–186.

Aslyng, H. C. (1963). Soil physics terminology. Int. Soc. Soil Sci. Bull. 23:7–10.

Bailey, G. W., and J. L. White (1970). Factors influencing the adsorption and movement of pesticides in soils. In Residue Reviews, vol. 32, F. A. Gunther and J. D. Gunther (eds.). Springer-Verlag, New York, pp. 29–92.

Barnhisel, R. I. (1977). Chlorites and hydroxy interlayered vermiculite and smectite. In Minerals in Soil Environments, J. B. Dixon, S. B. Weed, J. A. Kittrick, M. H. Milford, and J. L. White (eds.). Soil Sci. Soc. Am., Madison, Wis., pp. 331–356.

Baver, L. D. (1963). The effect of organic matter on soil structure. Pontif. Acad. Sci. Scr. Varia *32*:383–413.

Bernstein, L., and L. E. Francois (1973). Leaching requirement studies: Sensitivity of alfalfa to salinity of irrigation and drainage waters. Soil Sci. Soc. Am. Proc. *37*:931–943.

Besoain, E. (1968). Imogolite in volcanic soils of Chile. Geoderma *2*:151–169.

Birkeland, P. W. (1974). *Pedology, Weathering, and Geomorphological Research.* Oxford University Press, New York.

Bjerrum, N. (1923). Die Konstitution der Ampholyte, besonders der Aminosäuren, und ihre Dissoziations-Konstanten. Z. Phys. Chem. *104*:147–173.

Bloomfield, C. (1951). Experiments on the mechanism of gley formation. J. Soil Sci. *2*:196–211.

Bloomfield, C. (1953). A study of podzolization. Part 1. The mobilization of iron and aluminum by Scots pine needles. J. Soil Sci. *4*:5–16.

Bolt, G. H. (1967). Cation-exchange equations used in soil science—a review. Neth. J. Agric. Sci. *15*:81–103.

Bolt, G. H. (1976). Adsorption of anions by soils. In *Soil Chemistry. A. Basic Elements*, G. H. Bolt and M. G. M. Bruggenwert (eds.). Elsevier Scientific, Amsterdam, pp. 91–95.

Bolt, G. H., M. G. M. Bruggenwert, and A. Kamphorst (1976). Adsorption of cations by soil. In *Soil Chemistry. A. Basic Elements*, G. H. Bolt and M. G. M. Bruggenwert (eds.). Elsevier Scientific, Amsterdam, pp. 54–95.

Bowden, J. W., A. M. Posner, and J. P. Quirk (1977). Ionic adsorption on variable charge mineral surfaces. Theoretical charge development and titration curves. Aust. J. Soil Res. *15*:121–136.

Brady, N. C. (1974). *The Nature and Properties of Soils*, 8th ed. Macmillan, New York.

Brady, N. C. (1990). *The Nature and Properties of Soils*, 10th ed. MacMillan, New York.

Bragg, Sir L., and G. F. Claringbull (1965). *Crystal Structures of Minerals.* Bell, London.

Brewer, R., and J. R. Sleeman (1960). Soil structure and fabric. Their definition and description. J. Soil Sci. *11*:172–185.

Brindley, G. W., S. W. Bailey, G. T. Faust, S. A. Forman, and C. I. Rich (1968). Report of the nomenclature committee (1966–1967) of the Clay Minerals Society. Clays Clay Miner. *16*:322–324.

Brunauer, S., P. H. Emmett, and E. Teller (1938). Adsorption of gases in multimolecular layers. J. Am. Chem. Soc. *60*:309.

Buol, S. W., R. J. McCracken, and F. D. Hole (1973). *Soil Genesis and Classification.* Iowa State University Press, Ames.

Burdick, E. M. (1965). Commercial humates for agriculture and the fertilizer industry. Econ. Bot. *19*:152–156.

Burges, A. (1960). The nature and distribution of humic acid. Sci. Proc. R. Dublin Soc. Ser. A *1*:53–59.

Chapman, D. L. (1913). A contribution to the theory of electrocapillarity. Philos. Mag. *25*(6):475–481.

Chen, Y., and M. Schnitzer (1976). Scanning electron microscopy of a humic acid and its metal and clay complexes. Soil Sci. Soc. Am. J. *40*:682–686.

Chen, Y., N. Senesi, and M. Schnitzer (1977). Information provided on humic substances by E_4/E_6 ratios. Soil Sci. Soc. Am. J. *41*:352–358.

Chen, Y., N. Senesi, and M. Schnitzer (1978). Chemical and physical characteristics of humic and fulvic acids extracted from soils of the Mediterranean region. Geoderma *20*:87–104.

Clark, F. E., and K. H. Tan (1969). Identification of a polysaccharide ester linkage in humic acid. Soil Biol. Biochem. *1*:75–81.

Cragg, R. H. (1971). Lord Ernest Rutherford. Chem. Br. 7:518.

Cranwell, P. A., and R. D. Haworth (1975). The chemical nature of humic acids. In *Humic Substances, Their Structure and Function in the Biosphere*, D. Povoledo and H. L. Golterman (eds.). Centre Agric. Publ. Doc., Wageningen, Netherlands, pp. 13–18.

Davies, C. W. (1962). *Ion Association*. Butterworth, Washington, D.C.

Davis, L. E. (1945). Simple kinetic theory of ionic exchange for ions of unequal charge. J. Phys. Chem. *49*:473–479.

Degens, E. T., and K. Mopper (1975). Early diagenesis of organic matter in marine soils. Soil Sci. *119*:65–72.

Dixon, J. B. (1977). Kaolinite and serpentine group minerals. In *Minerals in Soil Environment*, J. B. Dixon, S. B. Weed, J. A. Kittrick, M. H. Milford, and J. L. White (eds.). Soil Sci. Soc. Am. Madison, Wis., pp. 357–403.

Donahue, R. L., R. W. Miller, and J. C. Shickluna (1977). *Soils, An Introduction to Soils and Plant Growth*. Prentice-Hall, Englewood Cliffs, N.J.

Dormaar, J. F. (1974). Scanning electron microscopy as applied to organomineral complexes in alkaline extracts of soil. Soil Sci. Soc. Am. Proc. *38*:685–686.

Dormaar, J. F. (1975). Effects of humic substances from Chernozemic A_h horizons on nutrient uptake by *Phaseolus vulgaris* and *Festuca sabrella*. Can. J. Soil Sci. *55*:111–118.

Douglas, L. A. (1977). Vermiculites. In *Minerals in Soil Environments*, J. B. Dixon, S. B. Weed, J. A. Kittrick, M. H. Milford, and J. L. White (eds.). Soil Sci. Soc. Am., Madison, Wis., pp. 259–292.

Egawa, T., and Y. Watanabe (1964). Electron micrographs of the clay minerals in Japanese soils. Bull. Natl. Inst. Agric. Sci. (Jpn.) Series *B 14*.

Eltantawy, I. M. (1980). The effect of heating on humic acid structure and electron spin resonance signal. Soil Sci. Soc. Am. J. *44*:512–514.

Eltantawy, I. M., and M. Baverez (1978). Structural study of humic acids by x-ray, electron spin resonance, and infrared spectroscopy. Soil Sci. Soc. Am. J. *42*:903–905.

Eriksson, E. (1952). Cation exchange equilibria on clay minerals. Soil Sci. *74*:103–113.

Eswaran, H. (1972). Morphology of allophane, imogolite and a halloysite. Clay Miner. *9*:281–285.

Eswaran, H., G. Stoops, and C. Sys (1977). The micromorphology of gibbsite forms in soils. J. Soil Sci. *28*:136–143.

Evans, R. C. (1939). *An Introduction to Crystal Chemistry*. Cambridge University Press, London.

Fanning, D. S., and V. Z. Keramidas (1977). Micas. In *Minerals in Soil Environments*, J. B. Dixon, S. B. Weed, J. A. Kittrick, M. H. Milford, and J. L. White (eds.). Soil Sci. Soc. Am., Madison, Wis., pp. 195–258.

Farmer, V. C. (1968). Infrared spectroscopy in clay mineral studies. Clay Miner. *7*:373–387.

Farmer, V. C., and F. Palmieri (1975). The characterization of soil minerals by infrared spectroscopy. In *Soil Components*, Vol. 2. *Inorganic Components*, J. E. Gieseking (ed.). Springer-Verlag, New York, pp. 573–671.

Farmer, V. C., A. R. Fraser, J. D. Russell, and N. Yoshinaga (1977). Recognition of imogolite structure in allophane clays by infrared spectroscopy. Clay Miner. *12*:55–57.

Farmer, V. C., W. J. McHardy, L. Robertson, A. Walker, and M. J. Wilson (1985). Micromorphology and sub-microscopy of allophane and imogolite in a Podzol B_s horizon: Evidence for translocation and origin. J. Soil Sci. *36*:87–95.

Felbeck, Jr., G. T. (1965). Structural chemistry of soil humic substances. Ad. Agron. *17*:327–368.

Fisher, M. J., D. A. Charles-Edwards, and M. M. Ludlow (1981). An analysis of the effects of repeated short-term soil water deficits on stomatal conductance to carbon dioxide and leaf photosynthesis by the legume *Macroptilium atropurpureum* cv. Siratro. Aust. J. Plant Physiol. *8*:347–357.

Flaig, W. (1975). An introductory review on humic substances: Aspects of research on their genesis, their physical and chemical properties, and their effect on organisms. In *Humic Substances, Their Structure and Function in the Biosphere*, D. Povoledo and H. L. Golterman (eds.). Centre Agric. Publ. Doc., Wageningen, Netherlands, pp. 19–42.

Flaig, W., and H. Beutelspacher (1951). Electron microscope investigations on natural and synthetic humic acids. Z. Pflanzenernahr. Dung. Bodenkd. *52*:1–21.

Flaig, W., H. Beutelspacher, and E. Rietz (1975). Chemical composition and physical properties of humic substances. In *Soil Components*. Vol. 1. *Organic Components*, J. E. Gieseking (ed.). Springer-Verlag, New York, pp. 1–211.

Foster, M. D. (1962). Interpretation of the composition and a classification of the chlorites. U. S. Geological Survey Professional Paper 414A.

Foth, H. D., and L. M. Turk (1978). *Fundamentals of Soil Science*. John Wiley & Sons, New York.

Freundlich, H. (1926). *Colloid and Capillary Chemistry*. Methuen, London.

Frissel, M. J., and G. H. Bolt (1962). Interaction between certain ionizable organic compounds (herbicides) and clay minerals. Soil Sci. *94*:284–291.

Gaillard, M. K. (1983). Toward a unified picture of elementary particle interactions. Am. Sci. *70*:506–514.

Gapon, E. N. (1933). Theory of exchange adsorption in soils. J. Gen. Chem. (USSR) *3*:144–152.

Garrels, R. M., and C. L. Christ (1965). *Solutions, Minerals, and Equilibria*. Harper & Row, New York.

Garrett, A. B. (1962). The flash genius: The Bohr atomic model: Niels Bohr. J. Chem. Educ. *39*:534.

Gast, R. C. (1977). Surface and colloid chemistry. In *Minerals in Soil Environments*, J. B. Dixon, S. B. Weed, J. A. Kittrick, M. H. Milford, and J. L. White (eds.). Soil Sci. Soc. Am., Madison, Wis., pp. 27–73.

Glashow, S. L. (1986). Neutrino. In *McGraw-Hill Yearbook of Science and Technology*. McGraw-Hill, New York.

Glasstone, S. (1946). *Textbook of Physical Chemistry*. Van Norstrand, Princeton, N.J.

Goenadi, D. H., and K. H. Tan (1988). Differences in clay mineralogy and oxidic ratio of selected LAC soils in temperate and tropical regions. Soil Sci. *146*:151–159.

Goenadi, D. H., and K. H. Tan (1991). The weathering of paracrystalline clays into kaolinite in Andisols and Ultisols in Indonesia. Indonesian J. Trop. Agric. *2*:56–65.

Goh, K. M., and F. J. Stevenson (1971). Comparison of infrared spectra of synthetic and natural humic and fulvic acids. Soil Sci. *112*:392–400.

Gortner, R. A., and W. A. Gortner (1949). *Outlines of Biochemistry*. John Wiley & Sons, New York.

Ghosh, K., and M. Schnitzer (1980a). Effects of pH and neutral electrolyte concentration on free radicals in humic substances. Soil Sci. Soc. Am. J. *44*:975–978.

Ghosh, K., and M. Schnitzer (1980b). Fluorescence excitation spectra of humic substances. Can. J. Soil Sci. *60*:373–379.

Ghosh, K., and M. Schnitzer (1980c). Macromolecular structures of humic substances. Soil Sci. *129*:266–276.

Gouy, G. (1910). Sur la constitution de la charge électrique à la surface d'un électrolyte. Ann. Phys. (Paris) Ser. 4 *9*:457–468.

Greenland, D. J. (1965). Interaction between clays and organic compounds in soils. 1. Mechanisms of interaction between clays and defined organic compounds. Soils Fertil. 28:415–425.

Greenland, D. J. (1971). Interactions between humic and fulvic acids and clays. Soil Sci. 111:34–41.

Greenland, D. J., and M. H. B. Hayes (eds.) (1978). The Chemistry of Soil Constituents. Wiley-Interscience, New York.

Greenland, D. J., G. R. Lindstrom, and J. P. Quirk (1961). Role of polysaccharides in stabilisation of natural soil aggregates. Nature 191:1283–1284.

Greenland, D. J., G. R. Lindstrom, and J. P. Quirk (1962). Organic materials which stabilize natural soil aggregates. Soil Sci. Soc. Am. Proc. 26:366–371.

Greenland, D. J., and J. P. Quirk (1962). Surface areas of soil colloids. Trans. Int. Soil Sci. Conf. Commun. IV and V, Palmerston North, New Zealand, pp. 79–87.

Greenland, D. J., and J. P. Quirk (1964). Determination of the total surface areas of soils by adsorption of cetyl pyridium bromide. J. Soil Sci. 15:178–191.

Grimshaw, R. W. (1971). The Chemistry and Physics of Clays. Wiley-Interscience, New York.

Guminski, S. (1957). The mechanism and conditions of the physiological actions of humic substances on the plant. Pochvovedenie 12:36.

Guminski, S., and Z. Guminska (1953). Studies on the activity of humus on plants. Acta Soc. Bot. Pol. 22:45–54.

Guminski, S., D. Augustin, and J. Sulej (1977). Comparison of some chemical and physico-chemical properties of natural and model sodium humates and of biological activity of both substances in tomato water cultures. Acta Soc. Bot. Pol. XIVI:437–448.

Hall, M. B. (1966). The background of Dalton's atomic theory. Chem. Br. 2:341.

Hatcher, P. G., R. Rowan, and M. A. Mattingly (1980). ^1H and ^{13}C NMR of marine humic acids. Org. Chem. 2:77–85.

Hem, J. D., and W. H. Cropper (1959). Survey of ferrous–ferric chemical equilibria and redox potentials. U.S. Geological Survey Water Supply Paper 1459A.

Hendricks, S. B. (1941). Base exchange of the clay mineral montmorillonite for organic cations and its dependence upon adsorption due to van der Waals forces. J. Phys. Chem. 45:65–81.

Hillel, D. (1972). Soil and Water. Physical Principles and Processes. Academic Press, New York.

Hodgson, J. F. (1969). Metal–organic complexing agents and transport of metal to roots. Soil Sci. Soc. Am. Proc. 33:68–75.

Hoffmann, R. W., and G. W. Brindley (1960). Adsorption of nonionic aliphatic molecules from aqueous solutions on montmorillonite. Clay–organic studies II. Geochim. Cosmochim. Acta *20*:15–29.

Holty, J. G., and P. E. Heilman (1971). Molecular sieve fractionation of organic matter in a Podzol from southeastern Alaska. Soil Sci. *112*:351–356.

Inoue, K., and P. M. Huang (1990). Perturbation of imogolite formation by humic substances. Soil Sci. Soc. Am. J. *54*:1490–1497.

Inoue, T., and K. Wada (1973). Adsorption of humified clover extracts by various clays. *Trans. 9th Int. Congr. Soil Sci.* Vol. 3, *Adelaide, Australia, 1968*. American Elsevier, New York, pp. 289–298.

Jackson, M. L., and G. D. Sherman (1953). Chemical weathering of minerals in soils. Adv. Agron. *5*:219–318.

Jackson, T. A. (1975). Humic matter in natural waters and sediments. Soil Sci. *119*:56–64.

Jeffery, J. W. O. (1960). Iron and the E_h of waterlogged soils with particular reference to paddy. J. Soil Sci. *11*:140–148.

Jenny, H. (1936). Simple kinetic theory of ionic exchange. I. Ions of equal valency. J. Phys. Chem. *40*:501–517.

Kamphorst, A., and G. H. Bolt (1976). Saline and sodic soils. In *Soil Chemistry*. A. *Basic Elements*, G. H. Bolt and M. G. M. Bruggenwert (eds.). Elsevier Scientific, Amsterdam, pp. 171–191.

Keller, W. D. (1954). Bonding energies of some silicate minerals. Am. Mineral. *39*:783–793.

Keller, W. D. (1964). Processes of origin and alteration of clay minerals. In *Soil Clay Mineralogy. A Symposium*, C. I. Rich and G. W. Kunze (eds.). University of North Carolina Press, Chapel Hill, pp. 3–76.

Kittrick, J. A. (1967). Gibbsite–kaolinite equilibria. Soil Sci. Soc. Am. Proc. *31*:314–316.

Kittrick, J. A. (1969). Soil minerals in the Al_2O_3–SiO_2–H_2O system and a theory of their formation. Clays Clay Miner. *17*:157–167.

Kleyn, W. B., and J. D. Oster (1983). Effects of permanent charge on the electric double layer properties of clays and oxides. Soil Sci. Soc. Am. J. *47*:821–827.

Klotz, I. M. (1950). *Chemical Thermodynamics*. Prentice-Hall, Englewood Cliffs, N.J.

Kolthoff, I. M., and E. B. Sandell (1952). *Textbook of Quantitative Inorganic Analysis*. Macmillan, New York.

Kononova, M. M. (1961). *Soil Organic Matter*, T. Z. Nowakowski and G. A. Greenwood (trans.). Pergamon, Oxford.

Kononova, M. M. (1966). *Soil Organic Matter*. Pergamon, Oxford, pp. 400–404.

Krauskopf, K. B. (1956). Dissolution and precipitation of silica at low temperatures. Geochim. Cosmochim. Acta *10*:1–26.

Krishnamoorty, C., and R. Overstreet (1949). Theory of ion exchange relationships. Soil Sci. *68*:307–315.

Kruyt, H. R. (1944). *Inleiding tot de Physische Chemie*. Uitgeverij H. J. Paris, Amsterdam.

Kubiena, W. L. (1938). *Micropedology*. Collegiate Press, Ames, Iowa.

Kumada, K. (1965). Studies on the colour of humic acids. Part 1. On the concepts of humic substances and humification. Soil Sci. Plant Nutr. *11*:151–156.

Kumada, K., and H. M. Hurst (1967). Green humic acid and its possible origin as a fungal metabolite. Nature *214*:5–88.

Kumada, K., and E. Miyara (1973). Sephadex gel filtration of humic acids. Soil Sci. Plant Nutr. *19*:255–263.

Lagerwerff, J. V., and G. H. Bolt (1959). Theoretical and experimental analysis of Gapon's equation for ion exchange. Soil Sci. *87*:127–222.

Langmuir, I. (1918). The adsorption of gases on plane surfaces of glass, mica and platinum. J. Am. Chem. Soc. *40*:1361–1382.

Lewis, G. N., and M. Randall (1921). The activity coefficient of strong electrolytes. J. Am. Chem. Soc. *43*:1112.

Lindsay, W. L. (1974). Role of chelation in micronutrient availability. In *The Plant Root and Its Environment*, E. W. Carson (ed.). University Press of Virginia, Charlottesville, pp. 507–524.

Lindsay, W. L., and W. A. Norvell (1969). Equilibrium relationships of Zn^{2+}, Fe^{3+}, Ca^{2+}, and H^+ with EDTA and DTPA in soils. Soil Sci. Soc. Am. Proc. *33*:62–68.

Lindsay, W. L., M. Peech, and J. S. Clark (1959). Solubility criteria for the existence of variscite in soils. Soil Sci. Soc. Am. Proc. *23*:357–360.

Lobartini, J. C., K. H. Tan, J. A. Rema, A. R. Gingle, C. Pape, and D. S. Himmelsbach (1992). The geochemical nature and agricultural importance of commercial humic matter. Science of Total Environment *113*: 1–15.

Lowe, L. E., and W.-C. Tsang (1970). Distribution of a green humic acid component in forest humus layers of British Columbia. Can. J. Soil Sci. *50*:456–457.

McEwan, D. M. C., and A. Ruiz-Amil (1975). Interstratified clay minerals. In *Soil Components*, Vol. 2, *Inorganic Components*, J. E. Gieseking (ed.). Springer-Verlag, New York, pp. 265–334.

Mackenzie, R. C. (1975). The classification of soil silicates and oxides. In *Soil Components*, Vol. 2, *Inorganic Components*, J. E. Gieseking (ed.). Springer-Verlag, New York, pp. 1–25.

Mackenzie, R. C., and S. Caillere (1975). The thermal characteristics of soil minerals and the use of these characteristics in the qualitative and quantitative determination of clay minerals in soils. In *Soil Components*, Vol.

2, *Inorganic Components*, J. E. Gieseking (ed.). Springer-Verlag, New York, pp. 529–571.

Manahan, S. E. (1975). *Environmental Chemistry*. Willard Grant Press, Boston.

Manov, G. G., R. G. Bates, W. J. Hamer, and S. F. Acree (1943). Values of the constants in the Debye–Hückel equation for activity coefficients. J. Am. Chem. Soc. *65*:1765–1767.

Martell, A. E., and M. Calvin (1952). *Chemistry of the Metal Chelate Compounds*. Prentice-Hall, New York.

Martin, A. E., and R. Reeve (1960a). Chemical studies of podzolized illuvial horizons. V. The flocculation of humus by ferric and ferrous iron and by nickel. J. Soil Sci. *11*:382–393.

Martin, A. E., and R. Reeve (1960b). Chemical studies of podzolic illuvial horizons. IV. The flocculation of humus by aluminum. J. Soil Sci. *11*:369–381.

Martin, J. H., W. H. Leonard, and D. L. Stamp (1976). *Principles of Field Crop Production*. MacMillan, New York.

Martin, F., C. Saiz-Jimenes, and A. Cert (1977). Pyrolysis–gas chromatography–mass spectrometry of soil humic fractions. I. The low boiling point compounds. Soil Sci. Soc. Am. J. *41*:1114–1118.

Martin, J. P., J. O. Ervin, and R. A. Shepherd (1966). Decomposition of the iron, aluminum, zinc and copper salts or complexes of some microbial and plant polysaccharides in soil. Soil Sci. Soc. Am. Proc. *30*:196–200.

Martin, J. P., K. Haider, and E. Bondietto (1975). Properties of model humic acids synthesized by phenoloxidase and autoxidation of phenols and other compounds formed by soil fungi. In *Humic Substances, Their Structure and Function in the Biosphere*, D. Povoledo and H. Golterman (eds.). Centre Agric. Publ. Doc., Wageningen, Netherlands, pp. 171–186.

McCarthy, J. S., B. E. Norum, and R. C. York (1983). *A long range plan for nuclear science*, Report of DOE/NSF Nuclear Science Advisory Committee. Proc. 12th Int. Conf. High Energy Accelerations, Batavia, Ill.

Mehta, N. C., P. Dubach, and H. Deuel (1963). Untersuchungen Uber die Molekular gewichts Verteilung von Huminstoffen durch Gelfiltration an Sephadex. Z. Pflanzenernahr. Dung. Bodenkd. *102*:128–137.

Mekaru, T., and G. Uehara (1972). Anion adsorption of ferruginous tropical soils. Soil Sci. Soc. Am. Proc. *36*:296–300.

Mellor, D. P. (1964). Historical background and fundamental concepts. In *Chelating Agents and Metal Chelates*, F. P. Dwyer and D. P. Mellor (eds.). Academic Press, New York, pp. 1–50.

Miller, R. H., and J. F. Wilkinson (1979). Nature of the organic coating on sand grains of nonwettable golf greens. Soil Sci. Soc. Am. Proc. *41*:1203–1204.

Millot, G. (1970). *Geology of Clays*. Springer-Verlag, New York.

Minson, D. J., and J. R. Wilson (1980). Comparative digestibility of tropical and temperate forage—a contrast between grasses and legumes. J. Aust. Inst. Agric. Sci. *46*:247–249.

Moore, J. T., and R. H. Loeppert (1987). Significance of potassium chloride pH of calcareous soils. Soil Sci. Soc. Am. J. *51*:908–912.

Mortenson, J. L. (1961). Physico-chemical properties of a soil polysaccharide. *Trans. 7th Int. Congr. Soil Sci. 1960*, Madison, Wisconsin. American Elsevier, New York, II:98–104.

Mortenson, J. L., D. M. Anderson, and J. L. White (1965). Infrared spectrometry. In *Methods of Soil Analysis*, Part 1, C. A. Black, D. D. Evans, J. L. White, L. E. Ensminger, and F. E. Clark (eds.). Agronomy Series No. 9. Am. Soc. Agron., Madison, Wis., pp. 743–770.

Mortland, M. M. (1970). Clay–organic complexes and interactions. Adv. Agron. *22*:75–117.

Murmann, R. K. (1964). *Inorganic Complex Compounds*. Holt–Reinhold, New York.

Norrish, K. (1954). Swelling of montmorillonite. Disc. Faraday Soc. *18*:120–134.

Novozamsky, I., J. Beek, and G. H. Bolt (1976). Chemical equilibria. In *Soil Chemistry. A. Basic Elements*, G. H. Bolt and M. G. M. Bruggenwert (eds.). Elsevier Scientific, Amsterdam, pp. 13–42.

Oden, S. (1919). Humic acids. Kolloidchem. Beih. *11*:75.

Ogner, G. (1979). The ^{13}C nuclear magnetic resonance spectrum of a methylated humic acid. Soil Biol. Biochem. *11*:105–108.

Ogner, G. (1980). Analysis of the carbohydrates of fulvic and humic acids as their partially methylated alditol acetates. Geoderma *23*:1–10.

Olness, A., and C. E. Clapp (1973). Occurrence of collapsed and expanded crystals in montmorillonite–dextran complexes. Clays Clay Miner. *21*:289–293.

Olness, A., and C. E. Clapp (1975). Influence of a polysaccharide structure on dextran adsorption by montmorillonite. Soil Biol. Biochem. *7*:113–118.

Orioli, G. A., and N. R. Curvetto (1980). Evaluation of extractants for soil humic substances. 1. Isotachophoretic studies. Plant Soil *55*:353–361.

Orlov, D. S., and N. L. Erosiceva (1967). *Zur frage der Wechselwirkung von Huminsauren mit einigen Metallkationen*. Vestnik-Moskovskogo Universiteta. Biologia, Pochvovedinie 1, pp. 98–105.

Orlov, D. S., Y. M. Ammosova, and G. I. Glebova (1975). Molecular parameters of humic acids. Geoderma *13*:211–229.

Orr, J. R. (1986). Particle accelerator. In *McGraw-Hill Yearbook of Science and Technology*. McGraw-Hill, New York.

Parfitt, R. L. (1992). Definitions of allophane. AIPEA (Assoc. Intern. pour l'Etude des Argilles) News Lett. *28*:16–19.

Parfitt, R. L., and J. M. Kimble (1989). Conditions for formation of allophane in soils. Soil Sci. Soc. Am. J. *53*:971–977.

Paton, T. R. (1978). *The Formation of Soil Material*. Allen & Unwin, Boston.

Paul, E. A., and F. E. Clark (1989). *Soil Microbiology and Biochemistry*. Academic Press, San Diego.

Pauling, L. (1929). The principles underlying the structure of complex ionic compounds. J. Am. Chem. Soc. *51*:1010–1026.

Perrin, D. D. (1964). *Organic Complexing Reagents*. Wiley–Interscience, New York.

Pierce, R. H., Jr., and G. T. Felbeck, Jr. (1975). A comparison of three methods of extracting organic matter from soils and marine sediments. In *Humic Substances, Their Structure and Function in the Biosphere*, D. Povoledo and H. L. Golterman (eds.). Centre Agric. Publ. Doc., Wageningen, Netherlands, pp. 217–232.

Poapst, P. A., C. Genier, and M. Schnitzer (1970). Effect of a soil fulvic acid on stem elongation in peas. Plant Soil *32*:367–372.

Poapst, P. A., C. Genier, and M. Shnitzer (1971). Fulvic acid and adventitious root formation. Soil Biol. Biochem. *3*:367–372.

Posner, A. M. (1964). Titration curves of humic acid. *Trans. Int. Congr. Soil Sci.*, Bucarest, Romania, 1964. Academy of the Socialist Republic of Romania, Bucharest, *11*:161–164.

Rich, C. I. (1968). Mineralogy of soil potassium. In *The Role of Potassium in Agriculture*, V. J. Kilmer, S. E. Younts, and N. C. Brady (eds.). Am. Soc. Agron., Madison, Wis., pp. 79–108.

Richards, L. A. (ed.). (1954). *Diagnosis and Improvement of Saline and Alkali Soils*. USDA Agriculture Handbook No. 60. U.S. Government Printing Office, Washington, D.C.

Riffaldi, R., and M. Schnitzer (1972). Electron spin resonance spectrometry of humic substances. Soil Sci. Soc. Am. Proc. *36*:301–307.

Ross, C. S., and P. F. Kerr (1934). Halloysite and allophane. U.S. Geological Survey Professional Paper 185G, pp. 135–148.

Ruggiero, P., F. S. Interesse, and O. Sciacovelli (1980). ^{1}H NMR evidence of exchangeable aromatic protons in fulvic and humic acids. Soil Biol. Biochem. *12*:297–299.

Salfeld, J. C. (1975). Ultraviolet and visible adsorption spectra of humic systems. In *Humic Substances. Their Structure and Function in the Biopshere*, D. Povoledo and H. L. Golterman (eds.). Centre Agric. Publ. Doc., Wageningen, Netherlands, pp. 269–280.

Schnitzer, M. (1965). The application of infrared spectroscopy to investigations on soil humic compounds. Can. Spectrosc. *10*:121–127.

Schnitzer, M. (1971). Characterization of humic constitutents by spectroscopy. In *Soil Biochemistry*, Vol. 2, A. D. McLaren and J. Skujins (eds.). Marcel Dekker, New York, pp. 60–95.

Schnitzer, M. (1974). The methylation of humic substances. Soil Sci. *117*:94–102.

Schnitzer, M. (1975). Chemical, spectroscopic and thermal methods for the classification and characterization of humic substances. In *Humic Substances, Their Structure and Function in the Biosphere*, D. Povoledo and H. L. Golterman (eds.). Centre Agric. Publ. Doc., Wageningen, Netherlands, pp. 293–310.

Schnitzer, M. (1976). The chemistry of humic substances. In *Environmental Biogeochemistry*. Vol. 1. *Carbon, Nitrogen, Phosphorus, Sulfur, and Selenium Cycles*, J. O. Nriagu (ed.). Proc. 2d Int. Symp. Environ. Biogeochem., Hamilton, Ontario, Canada, April 8–12, 1975. Ann Arbor Science, Ann Arbor, Mich., pp. 89–107.

Schnitzer, M., and S. U. Khan (1972). *Humic Substances in the Environment*. Marcel Dekker, New York.

Schnitzer, M., and H. Kodama (1976). The dissolution of micas by fulvic acid. Geoderma *15*:381–391.

Schnitzer, M., and D. A. Hindle (1980). Effect of peracetic acid oxidation on N-containing components of humic materials. Can. J. Soil Sci. *60*:541–548.

Schnitzer, M., D. A. Shearer, and J. R. Wright (1959). A study in the infrared of high-molecular-weight organic matter extracted by various reagents from a Podzolic B horizon. Soil Sci. *87*:252–257.

Schofield, R. K. (1947). A ratio law governing the equilibrium of cations in the soil solution. Proc. 11th Int. Congr. Pure Appl. Chem. *3*:257–261.

Schofield, R. K. (1955). Can a precise meaning be given to "available phosphorus"? Soils Fertil. *18*:373–375.

Schofield, R. K., and A. W. Taylor (1955). The mesurement of soil pH. Soil Sci. Soc. Am. Proc. *19*:164–167.

Schwertmann, U., and R. M. Taylor (1977). Iron oxides. In *Minerals in Soil Environments*, J. B. Dixon, S. B. Weed, J. A. Kittrick, M. H. Milford, and J. L. White (eds.), Soil Sci. Soc. Am. Madison, Wis., pp. 145–180.

Senesi, N., and M. Schnitzer (1977). Effects of pH, reaction time, chemical reduction and irradiation on ESR spectra of fulvic acid. Soil Sci. *123*:224–234.

Simonson, R. W. (1959). Outline of a generalized theory of soil genesis. Soil Sci. Soc. Am. Proc. *23*:152–156.

Slatyer, R. O. (1957). The significance of the permanent wilting percentage in studies of plant and soil water relations. Bot. Rev. *23*:585–636.

Slatyer, R. O. (1967). *Plant–Water Relationships*. Academic Press, New York.

Sposito, G. (1980). Freundlich equation for ion exchange reactions in soils. Soil Sci. Soc. Am. J. *44*:652–654.

Sposito, G., and K. M. Holtzclaw (1977). Titration studies on the polynuclear, polyacidic nature of fulvic acid extracted from sewage sludge–soil mixtures. Soil Sci. Soc. Am. J. *41*:330–336.

Steelink, C. (1964). Free radical studies of lignin, lignin degradation products and soil humic acids. Geochim. Cosmochim. Acta *28*:1615–1622.

Steelink, C., and G. Tollin (1967). Free radicals in soils. In *Soil Biochemistry*, A. D. McLaren and G. H. Peterson (eds.). Marcel Dekker, New York, pp. 147–173.

Stern, O. (1924). Zür Theory der elektrolytischen Doppelschicht. Z. Elektrochem. *30*:508–516.

Stevenson, F. J. (1965). Gross chemical fractionation of organic matter. In *Methods of Soil Analysis*, Part 2, C. A. Black, D. D. Evans, J. L. White, L. E. Ensminger, and F. E. Clark (eds.). Agron. Ser. 9. Am. Soc. Agron., Madison, Wis., pp. 1409–1421.

Stevenson, F. J. (1976a). Stability constants of Cu^{2+}, Pb^{2+} and Cd^{2+} complexes with humic acids. Soil Sci. Soc. Am. J. *40*:665–672.

Stevenson, F. J. (1976b). Binding of metal ions by humic acids. In *Environmental Biogeochemistry*. Vol. 1. *Carbon, Nitrogen, Phosphorus, Sulfur, and Selenium Cycles*. J. O. Nriagu (ed.). Proc. 2d. Int. Symp. Environ. Biogeochem., Hamilton, Ontario, Canada, April 8–12, 1975. Ann Arbor Science, Ann Arbor, Mich., pp. 519–540.

Stevenson, F. J., and K. M. Goh (1971). Infrared spectra of humic acids and related substances. Geochim. Cosmochim. Acta *35*:471–483.

Stevenson, F. J., and K. M. Goh (1972). Infrared spectra of humic and fulvic acid and their methylated derivatives, evidence for nonspecificity of analytical methods for oxygen-containing functional groups. Soil Sci. *113*:L334–345.

Stevenson, I. L., and M. Schnitzer (1982). Transmission electron microscopy of extracted fulvic and humic acids. Soil Sci. *133*:179–185.

Sticher, H., and R. Bach (1966). Fundamentals in the chemical weathering of silicates. Soils Fertil. *29*:321–325.

Sudo, T., and H. Yotsumoto (1977). The formation of halloysite tubes from spherulitic halloysite. Clays Clay Miner. *25*:155–159.

Swift, R. S., B. K. Thornton, and A. M. Posner (1970). Spectral characteristics of a humic acid fractionated with respect to molecular weight using agar gel. Soil Sci. *110*:93–99.

Swindale, L. D. (1975). The crystallography of minerals of the kaolin group. In *Soil Components*, Vol. 2, *Inorganic Components*, J. E. Gieseking (ed.). Springer-Verlag, New York, pp. 121–154.

Tait, J. M., N. Yoshinaga, and B. D. Mitchell (1978). Occurrence of imogolite in some Scottish soils. Soil Sci. Plant Nutr. *24*:145–151.

Tan, K. H. (1964). The Andosols in Indonesia. In *F.A.O. Meeting on the Classification and Correlation of Soils from Volcanic Ash*. World Soil Resources. F.A.O. Rep. 14. U.N. Educ. Sci. Cult. Org., Rome, Italy, pp. 23–30.

Tan, K. H. (1975). Infrared absorption similarities between hymatomelanic acid and methylated humic acid. Soil Sci. Soc. Am. Proc. *39*:70–73.

Tan, K. H. (1976a). Complex formation between humic acid and clays as revealed by gel filtration and infrared spectroscopy. Soil Biol. Biochem. *8*:235–239.

Tan, K. H. (1976b). Contamination of humic acid by silica gel and sodium bicarbonate. Plant Soil *44*:691–695.

Tan, K. H. (1978a). Formation of metal–humic acid complexes by titration and their characterization by differential thermal analysis and infrared spectroscopy. Soil Biol. Biochem. *10*:123–129.

Tan, K. H. (1978b). Effect of humic and fulvic acids on release of fixed potassium. Geoderma *21*:67–74.

Tan, K. H. (1978c). Variations in soil humic compounds as related to regional and analytical differences. Soil Sci. *125*:351–358.

Tan, K. H. (1985). Scanning electron microscopy of humic matter as influenced by methods of preparation. Soil Sci. Soc. Am. J. *49*:1185–1191.

Tan, K. H. (1986). Degradation of soil minerals by organic acids. In *Interactions of Soil Minerals with natural Organics and Microbes*, P. M. Huang and M. Schnitzer (eds.). SSSA Spec. Publ. 17. Soil Sci. Soc. Am., Madison, Wis.

Tan, K. H., and A. Binger (1986). Effect of humic acid on aluminum toxicity in corn plants. Soil Sci. *141*:20–25.

Tan, K. H., and J. Van Schuylenborgh (1959). On the classification and genesis of soils derived from andesitic volcanic material under a monsoon climate. Neth. J. Agric. Sci. 7:1–22.

Tan, K. H., and J. Van Schuylenborgh (1961). On the organic matter in tropical soils. Neth. J. Agric. Sci. 9:174–180.

Tan, K. H., and F. E. Clark (1968). Polysaccharide constituents in fulvic and humic acids extracted in soils. Geoderma 2:245–255.

Tan, K. H., and R. A. McCreery (1970a). Possibility of the silylation technique in gas liquid chromotography of fulvic and humic acids. Geoderma *4*:119–126.

Tan, K. H., and R. A. McCreery (1970b). The infrared identification of a humo–polysaccharide ester in soil humic acid. Commun. Soil Sci. Plant Anal. *1*:75–84.

Tan, K. H., and J. E. Giddens (1972). Molecular weights and spectral characteristics of humic and fulvic acids. Geoderma 8:221–229.

Tan, K. H., and R. A. McCreery (1975). Humic acid complex formation and intermicellar adsorption by bentonite. Proc. 1974 Int. Clay Conf., Mexico City, Mexico. Applied, Wilmette, Ill., pp. 629–641.

Tan, K. H., and B. F. Hajek (1977). Thermal analysis of soils. In *Minerals and Soil Environments*, J. B. Dixon, S. B. Weed, J. A. Kittrick, M. H. Milford, and J. L. White (eds.). Soil Sci. Soc. Am., Madison, Wis., pp. 865–884.

Tan, K. H., and V. Nopamornbodi (1979). Effect of different levels of humic acids on nutrient content and growth of corn (*Zea mays* L.). Plant Soil *51*:283–387.

Tan, K. H., and H. F. Perkins (1980). The value of the U.S. Soil Taxonomy classification of red tropical soils derived from volcanic ash. Proc. CLA-MATROPS, Int. Conf. Classif. Manage. Trop. Soils, Kuala Lumpur, Malaysia, Aug. 1977. Int. Soil Sci. Soc., Kuala Lumpur, Malaysia, pp. 110–119.

Tan, K. H., L. D. King, and H. D. Morris (1971a). Complex reactions of zinc with organic matter extracted from sewage sludge. Soil Sci. Soc. Am. Proc. *35*:748–751.

Tan, K. H., R. A. Leonard, A. R. Bertrand, and S. R. Wilkinson (1971b). The metal complexing capacity and the nature of the chelating ligands of water extract of poultry litter. Soil Sci. Soc. Am. Proc. *35*:266–269.

Tan, K. H., H. F. Perkins, and R. A. McCreery (1973). Kaolinite–gibbsite thermodynamic relationship in Ultisols. Soil Sci. *116*:8–12.

Tan, K. H., V. G. Mudgal, and R. A. Leonard (1975a). Adsorption of poultry litter extracts by soil and clay. Environ. Sci. Technol. *9*:132–135.

Tan, K. H., E. R. Beaty, R. A. McCreery, and J. B. Jones (1975b). Differential effect of Bermuda and Bahia grasses on soil chemical characteristics. Agron. J. *67*:407–411.

Tan, K. H., G. W. Bailey, and H. F. Perkins (1978). Infrared Analysis. In *Analysis of Clay, Silt and Sand Fractions of Selected Soils From the Southeastern United States*, Southern Coop. Ser. Bull. 219. University Kentucky, Lexington, Ky., pp. 25–38.

Taylor, S. A., and G. L. Ashcroft (1972). *Physical Edaphology*. Freeman, San Franciso.

Taylor, S. A., and R. O. Slatyer (1960). Water–soil plant relations terminology. Trans. 7th Int. Congr. Soil Sci., Madison, Wis., *1*:80–90.

Theng, B. K. G. (1974). The chemistry of clay–organic reactions. John Wiley & Sons, New York.

Theng, B. K. G. (1972). Formation, properties, and practical applications of clay–organic complexes. J. R. Soc. N. Z. *2*:437–457.

Thomas, G. W. (1974). Chemical reactions controlling soil solution electrolyte concentration. In *Plant Root and Its Environment*, E. W. Carson (ed.). University Press of Virginia, Charlottesville, pp. 483–506.

Tisdale, S. L., and W. L. Nelson (1975). *Soil Fertility and Fertilizers*. Macmillan, New York.

Tiurin, I. V., and M. M. Kononova (1962). Biology of humus formation and questions of soil fertility. Trans. Joint Meet. Commun. IV and V, Int. Soc. Soil Sci., Palmerston-North, New Zealand, pp. 203–219.

Tucker, K. A., K. J. Karnok, D. E. Radcliffe, G. Landry, Jr., R. W. Roncadory, and K. H. Tan (1990). Localized dry spots as caused by hydrophobic sands on bentgrass greens. Agron. J. *82*:549–555.

Turner, N. C., and M. M. Jones (1980). Turgor maintenance by osmotic adjustment. A review and evaluation. In *Adaptation of Plants to Water and High Temperature Stress*, N. C. Turner and P. J. Kramer (eds.). Wiley-Interscience, New York, pp. 87–103.

Turner, N. C., and P. J. Kramer (eds.). (1980). *Adaptation of Plants to Water and High Temperature Stress*. Wiley-Interscience, New York.

USDA Soil Conservation Service, Soil Survey Staff (1960). Soil classification, a comprehensive system, 7th approximation. U.S. Government Printing Office, Washington, D.C.

USDA Soil Conservation Service, Soil Survey Staff (1975). *Soil Taxonomy— A Basic System of Soil Classification for Making and Interpreting Soil Surveys*. Agriculture Handbook 436, USDA, SCS, U.S. Government Printing Office, Washington, D.C.

Van Breemen, N., and R. Brinkman (1976). Chemical equilibria and soil formation. In *Soil Chemistry*, A. *Basic Elements*. G. H. Bolt and M. G. M. Bruggenwert (eds.). Elsevier Scientific, Amsterdam, pp. 141–170.

Van der Marel, H. W. (1959). Potassium fixation, a beneficial soil characteristic for crop production. Z. Pflanzenernähr. Dung. Bodenkd. *84*:51–62.

Van Olphen, H. (1977). Clay colloid chemistry. John Wiley & Sons, New York.

Van Raij, B., and M. Peech (1972). Electrochemical properties of some Oxisols and Alfisols of the tropics. Soil Sci. Soc. Am. Proc. *36*:587–593.

Van Schuylenborgh, J. (1965). The formation of sesquioxides in soils. In *Experimental Pedology*, E. G. Hallsworth and D. V. Crawford (eds.). Butterworth, London, pp. 113–125.

Van Schuylenborgh, J. (1966). Chemical aspects of soil formation, syllabus of lectures postgraduate training. Agriculture University, Wageningen, Netherlands.

Van Schuylenborgh, J. (1971). Weathering and soil forming processes in the tropics. In *Proc. Bandung Symp. Soils Trop. Weathering, Indonesia*. Unesco, Paris, pp. 39–50.

Verwey, E. J. W., and J. T. G. Overbeek (1948). *Theory of the Stability of Lyophobic Colloids*. Elsevier, New York.

Wada, K. (1977). Allophane and imogolite. In *Minerals in Soil Environments*, J. B. Dixon, S. B. Weed, J. A. Kittrick, M. H. Milford, and J. L. White (eds.). Soil Sci. Soc. Am., Madison, Wis., pp. 603–639.

Walker, G. F. (1961). Vermiculite minerals. In *The X-ray Identification and Crystal Structures of Clay Minerals*. G. Brown (ed.). Mineralogical Society, London, pp. 297–324.

Walker, G. F. (1975). Vermiculites. In *Soil Components*, Vol. 2, *Inorganic Components*, J. E. Gieseking (ed.). Springer-Verlag, New York, pp. 155–189.

Wallace, J. M., and L. C. Whitehand (1980). Adverse synergistic effects between acetic, propionic, butyric, and valeric acids on the growth of wheat seedling roots. Soil Biol. Biochem. *12*:445–446.

Way, J. T. (1850). On the power of soils to absorb manure. J. R. Agric. Soc. *11*:313–379.

Weast, R. C. (ed.). (1972). *Handbook of Chemistry and Physics*. The Chemical Rubber Co., Cleveland, Ohio.

Weber, J. B. (1970). Mechanisms of adsorption of s-triazines by clay colloids and factors affecting plant availability. In *Residue Reviews*, Vol. 32, F. A. Gunther and J. D. Gunther (eds.), Springer-Verlag, New York, pp. 93–130.

White, J. L. (1971). Interpretation of infrared spectra of soil minerals. Soil Sci. *112*:22–31.

Wilkinson, J. P., and R. H. Miller (1978). Investigation and treatment of localized dry spots on sand golf greens. Agron. J. *70*:299–304.

Yates, D. E., S. Levine, and T. W. Healy (1974). Site-binding model of the electric double layer at the oxide/water interface. J. Chem. Soc. Faraday Trans. I *70*:1807–1819.

Yoshida, M., K. Sakagami, R. Hamada, and T. Kurobe (1978). Studies on the properties of organic matter in buried humic horizon derived from volcanic ash. I. Humus composition of buried humic horizon. Soil Sci. Plant Nutr. *24*:277–287.

Yoshinaga, N., and S. Aomine (1962). Imogolite in some Ando soils. Soil Sci. Plant Nutr. *8*:6–13.

Index